Nuclear Reactors - Spacecraft Propulsion, Research Reactors, and Reactor Analysis Topics

Edited by Chad L. Pope

Published in London, United Kingdom

IntechOpen

Supporting open minds since 2005

Nuclear Reactors - Spacecraft Propulsion, Research Reactors, and Reactor Analysis Topics
http://dx.doi.org/10.5772/intechopen.95676
Edited by Chad L. Pope

Contributors

Shannon L. Eggers, Robert Anderson, Alok Jha, Vinjay Kumar, Mohan Rao Mamdikar, Pooja Singh, Nicolas Woolstenhulme, Mark D. David DeHart, Sebastian Schunert, Vincent M. Labouré, Samuel E. Bays, Joseph W. Nielsen, Chad L. Pope, Edward Lum, Ryan Stewart, William Phoenix, Enrico Zio, Ibrahim Ahmed, Gyungyung Heo

Notice
Statements and opinions expressed in the chapters are these of the individual contributors and not necessarily those of the editors or publisher. No responsibility is accepted for the accuracy of information contained in the published chapters. The publisher assumes no responsibility for any damage or injury to persons or property arising out of the use of any materials, instructions, methods or ideas contained in the book.

First published in London, United Kingdom, 2022 by IntechOpen
IntechOpen is the global imprint of INTECHOPEN LIMITED, registered in England and Wales, registration number: 11086078, 5 Princes Gate Court, London, SW7 2QJ, United Kingdom
Printed in Croatia

British Library Cataloguing-in-Publication Data
A catalogue record for this book is available from the British Library

Additional hard and PDF copies can be obtained from orders@intechopen.com

Nuclear Reactors - Spacecraft Propulsion, Research Reactors, and Reactor Analysis Topics
Edited by Chad L. Pope
p. cm.
Print ISBN 978-1-83969-939-9
Online ISBN 978-1-83969-940-5
eBook (PDF) ISBN 978-1-83969-941-2

We are IntechOpen,
the world's leading publisher of
Open Access books
Built by scientists, for scientists

5,900+
Open access books available

146,000+
International authors and editors

185M+
Downloads

Our authors are among the

156
Countries delivered to

Top 1%
most cited scientists

12.2%
Contributors from top 500 universities

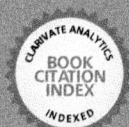

Interested in publishing with us?
Contact book.department@intechopen.com

Numbers displayed above are based on latest data collected.
For more information visit www.intechopen.com

Meet the editor

Chad L. Pope received his BS in General Engineering in 1989, MS in Nuclear Science and Engineering in 1993, and Ph.D. in Nuclear Science and Engineering in 2011, all from Idaho State University, USA. From 1989 to 2013, he worked for the Naval Reactors Facility, Argonne National Laboratory, and Idaho National Laboratory in nuclear safety while supporting the Experimental Breeder Reactor-II, the Neutron Radiography Reactor, the Transient Reactor Test Facility, and the Idaho State University AGN-201 nuclear reactor. Following a 24-year industry career, Dr. Pope accepted a position in the Idaho State University Nuclear Engineering Department in 2013 where he is engaged in nuclear engineering teaching and research, serves as the department chair, and is jointly appointed to Idaho National Laboratory.

Contents

Preface

Worldwide interest in nuclear reactors continues to increase and significant focus has been placed on advanced nuclear reactors intended to produce electricity and process heat. Somewhat absent from the broader discussion has been the importance of research reactors and certain specialized reactor analysis topics. This book attempts to fill this gap through three sections: "Nuclear Reactors for Spacecraft Propulsion", "Research Reactors", and "Select Reactor Analysis Techniques".

The first section of the book addresses the use of nuclear reactors for spacecraft propulsion. A detailed explanation is provided regarding the optimum approach for using a reactor as the heat source along with the basis for selecting hydrogen as the propulsion gas.

The second section of the book provides information about two landmark research reactors as well as a university reactor. The first research reactor discussed is the Transient Reactor Test Facility (TREAT) located in Idaho, USA. The TREAT reactor is an air-cooled graphite-moderated reactor used for testing new reactor fuel under extreme power transients. The TREAT reactor is capable of 18,000 MW power transients, which allows new fuel and material designs to be tested under severe accident conditions. The second research reactor discussed is the Experimental Breeder Reactor II (EBR II). EBR-II was a sodium-cooled fast-neutron-spectrum reactor that operated in Idaho, USA from 1964 to 1994. EBR-II provided comprehensive sodium-cooled fast reactor testing and operational experience. The most significant experiments conducted in EBR-II occurred in 1986 when the reactor was subjected to loss of flow and loss of heat sink, both without reactor SCRAM. In both experiments, the inherent safety properties of the reactor led to a safe shutdown without fuel damage, and no active engineered safety feature or operator action was required. The final reactor discussed is the Idaho State University AGN-201. This reactor has been operating for more than 50 years. The simple reactor design consists of UO_2 dispersed in polyethylene along with a graphite reflector. The reactor operates at a very low power of 5 W and requires no active cooling, making it ideal for a university teaching setting where physics can be studied without the burden of numerous operational constraints.

The third section of the book includes a collection of interesting reactor analysis topics. It provides a detailed discussion of the Advanced Test Reactor (ATR) core reloading analysis. The ATR is in Idaho, USA, and is the largest test reactor operated by the US government and has been operating for more than 50 years. The very high neutron flux (greater than 10^{14} neutrons/cm^2 s) in ATR necessitates frequent refueling but allows fuel and material testing to simulate years of reactor service in a matter of months. Additional reactor analysis topics discussed in this section include cyber-informed engineering for nuclear reactor digital instrumentation and control, a nuclear power plant case study focusing on instrumentation and control system reliability analysis, determining the plenum gas effect on fuel temperature, and finally, fault detection by signal reconstruction in nuclear power plants.

It is hoped that this collection of discussions on spacecraft nuclear reactor propulsion, research and university reactors, and various reactor analysis techniques will provide readers with valuable insights into important aspects of nuclear reactors that have not been well disseminated previously.

Chad L. Pope
Department of Nuclear Engineering,
Idaho State University,
Pocatello, USA

Section 1

Nuclear Reactors for Spacecraft Propulsion

Chapter 1

Nuclear Thermal Propulsion

Mark D. DeHart, Sebastian Schunert and Vincent M. Labouré

Abstract

This chapter will cover the fundamentals of nuclear thermal propulsion systems, covering basic principles of operation and why nuclear is a superior option to chemical rockets for interplanetary travel. It will begin with a historical overview from early efforts in the early 1950s up to current interests, with respect to fuel types, core materials, and ongoing testing efforts. An overview will be provided of reactor types and design elements for reactor concepts or testing systems for nuclear thermal propulsion, followed by a discussion of nuclear thermal design concepts. A section on system design and modeling will be presented to discuss modeling and simulation of driving phenomena: neutronics, materials performance, heat transfer, and structural mechanics, solved in a tightly coupled multiphysics system. Finally, it will show the results of a coupled physics model for a conceptual design with simulation of rapid startup transients needed to maximize hydrogen efficiency.

Keywords: neutronics, high temperature, multiphysics, griffin, MOOSE, nuclear thermal propulsion, interplanetary, NASA

1. Introduction

Nuclear thermal propulsion (NTP) is a technology that uses a nuclear reactor to provide the necessary energy to power a spacecraft for extraterrestrial operations [1]. At the most basic level, nuclear thermal propulsion is simply the use of nuclear fission to heat a gas to a high exit velocity. In this sense, it is very similar to a chemical rocket, in which the exothermic reaction of hydrogen and oxygen provides the energy used to heat the reaction product—gaseous H_2O—to generate thrust. However, in an NTP engine, molecular hydrogen (H_2) is used as the propellant. The H_2 is used to remove heat from a reactor core by convection; the added energy provides a high speed exit velocity to generate thrust.

For an NTP engine using an H_2 propellant, the engine is two to three times as efficient as an H_2/O_2-fueled rocket engine. Here, efficiency is measured in terms of *specific impulse* (I_{sp}). The Isp is the amount of time (in seconds) that a rocket engine can generate thrust with a fixed Earth weight (mass $\times g_o$) of propellant, when the weight is equal to the engine's thrust [2]. Here g_o is the gravitational constant on Earth, about 9.81 m/s^2, and relates mass to weight. For an H_2/O_2 engine, the I_{sp} is around 450 s. For nuclear thermal propulsion with H_2, the I_{sp} is approximately 900 s [3]. Hence, the United States (U.S.) National Aeronautics and Space Administration (NASA) has had a long interest in use of NTP for propulsion, with recent interest in missions to Mars between 2030 and 2050 [4], and for cislunar operations, with a plan to demonstrate an NTP system above low Earth orbit (LEO) by 2025 [5].

This chapter is organized as follows. First, background will be provided on historical NTP work and current needs for operation—specifically, the functionality of an NTP engine. Next, we will detail key components of core physics design, focusing on the nuclear subsystem of the larger plant. We will briefly discuss the balance of plant as it relates to the nuclear subsystem, then conclude with a presentation of simulation results for a conceptual nuclear thermal propulsion system.

2. Background

In this section, we will discuss the evolution of the NTP concept from its theoretical beginnings in the late 1940s to present-day needs. Much of the knowledge being applied in current NTP system design is drawn from knowledge gained through a series of experimental programs beginning in the late 1950s and running through the early 1970s. However, current interests have resulted in new materials testing based on experience gleaned from earlier work combined with modern materials performance data and testing methods. A major focus of the NASA Space Nuclear Propulsion Program is in reviving NTP fuel fabrication techniques and design knowledge [6]. Hence, an overview of the history of NTP is appropriate before moving on to current testing programs. These programs provide significant insight for current research and testing programs. But first, let us revisit the motivation for nuclear thermal propulsion over chemical engines for extraterrestrial propulsion.

2.1 Advantages of nuclear thermal propulsion for interplanetary travel

The efficiency of a rocket engine design is commonly measured in terms of specific impulse. One can think of I_{sp} as the miles per gallon or kilometers per liter for a car. The larger the I_{sp}, the more efficient the engine. Mathematically, specific impulse is defined as the total engine thrust integrated over time per unit weight of the propellant; here, weight is defined as measured on Earth (e.g., N, or, historically, lb$_f$) [7]). Thrust is defined as:

$$F_{thrust} = v_e \cdot \dot{m} \tag{1}$$

where:
F_{thrust} is the force (thrust) exerted by the propellant (N),
v_e is the exit velocity of the exhaust propellant (m/s) relative to the nozzle, and
$\dot{m}$, or dm/dt, is the mass flow rate of the propellant (kg/s).
The total impulse (I) of a rocket for time t is defined as the thrust integrated over the total time of operation (*burn time* in a chemical rocket, or time at power in an NTP engine):

$$I(t) = \int_0^t F_{thrust}(\tau)d\tau = \int_0^t v_e \cdot \dot{m} \, d\tau = m_{ex} \cdot v_e \tag{2}$$

Here, we have assumed that v_e is constant, and m_{ex} is the total mass expelled over the time of operation, $m(0) - m(t)$.
Specific impulse is defined as the total impulse divided by the weight W of the propellant on Earth, i.e.:

$$W = m_{ex} \cdot g_o \tag{3}$$

Hence,

$$I_{sp} = \frac{m_{ex} \cdot v_e}{W} = \frac{m_{ex} \cdot v_e}{m_{ex} \cdot g_o} = \frac{v_e}{g_o} \tag{4}$$

Rearranging the expressions in Eq. (1), in terms of v_e and replacing v_e in Eq. (4), we arrive at a more useful definition for I_{sp}:

$$I_{sp} = \frac{F_{thrust}}{\dot{m} \cdot g_o} \tag{5}$$

Eq. (5) shows that the I_{sp} is the ratio of thrust to the product of the mass flow rate times the constant g_o. In this form, it is clear that the I_{sp} can be interpreted as the time (in s) over which 9.81 kg (or one Newton of weight on Earth) of propellant can produce one Newton of thrust. The larger the I_{sp}, the longer the engine is able to operate with a given mass of fuel.

A pioneer in rocketry theory in the early 1890s, Konstantin Tsiolkovsky [8] derived a number of important relationships, including Eq. (6), which is used heavily in rocket design and is known as the ideal exhaust velocity equation, relating gas properties to the exit velocity of the propellant:

$$v_e^2 = \frac{\frac{2kRT_c}{M}\left(1 - \left(\frac{p_e}{p_c}\right)^{\frac{k-1}{k}}\right)}{k-1}, \tag{6}$$

where:

k is the ratio of the specific heat at constant pressure (c_p) to specific heat at constant volume (c_v) for the propellant (i.e., $k = c_p/c_v$),

R is the universal gas constant,

T_c is the reactor core exit temperature for NTP, or the combustion chamber temperature for a chemical engine,

M is the molecular weight of the propellant,

p_e is the nozzle exit pressure, and

p_c is the core exit (or combustion chamber) pressure.

R is a fundamental physical constant, k does not vary significantly between different gases (typically between 1.1 and 1.5) and T_c, p_c, and p_e depend on the engine specifications. Assuming that k, Tc, pc, and pe are known and identical between NTP and a chemical rocket, we can combine them into the constant C:

$$I_{sp} = \frac{C}{\sqrt{M}}, \tag{7}$$

For rockets that use the chemical reaction of H_2 and O_2 to produce energy and release high temperature H_2O, the atomic mass of the propellant, M, is 18 g/mole. NTP engines use high energy H_2 ($M = 2$ g/mole) that is discharged from a high temperature core. Comparing the theoretical specific impulses,

$$\frac{I_{sp}(H_2)}{I_{sp}(H_2O)} = \frac{C_{H_2O}/\sqrt{2}}{C_{H_2}/\sqrt{18}} = \sqrt{18/2} = 3. \tag{8}$$

This assumes that C_{H_2O} is equal to C_{H_2} (they are similar but not equal). Thus, based on ideal gas assumptions, H_2 could provide three times the I_{sp} of H_2O as a propellant. However, in reality, gas is not ideal and the value of C_{H_2O} is not equal to

C_{H_2}, as the value of k is not the same for the two fluids. In addition, for NTP, the most significant challenge is in obtaining a high exit temperature from the core. This requires nuclear fuel materials to be able to quickly rise to and maintain very high temperatures. Chemical engine combustion chamber temperatures are on the order of 3500 K; NTP efforts currently aim for a temperature of approximately 2700–3000 K based on material limits. Together, these facts somewhat reduce the advantage in I_{sp} from the ideal value of 3 to a ratio closer to 2. Nevertheless, with the I_{sp} for a H_2/O_2 engine is on the order of 450 s, while for NTP, it would be on the order of 900 s. Hence, there remains a clear advantage to the use of an NTP engine. Heating H_2 to significant outlet temperatures can be achieved using a nuclear reactor.

This advantage was recognized in the 1940s. An NTP-propelled spacecraft could significantly reduce the travel time to Mars as compared to conventional engines [9]. This would reduce astronaut radiation exposure, as well as the impact of the long-term microgravity environment.

Note that NTP engines are not intended for liftoff from Earth; they are not designed to provide sufficient thrust for launch. Chemical engines would be used to lift a full vessel (in parts) to low earth orbit (LEO), from where the vessel would be assembled and an NTP-propelled mission would be launched.

In the late 1960s, the well-known pioneer of modern rocketry, Wernher von Braun, then the director of the NASA Marshall Space Flight Center, advocated for a mission to Mars. Under his plan, NASA would launch a Mars mission in November 1981 (based on favorable planetary alignment), and land on the red planet by August 1982. Von Braun explained that "although the undertaking of this mission will be a great national challenge, it represents no greater challenge than the commitment made in 1961 to land a man on the moon" [10]. In the following subsection, we will briefly visit early NTP research and the Nuclear Engine for Rocket Vehicle Application (NERVA) rocket engines that von Braun had envisioned would take men to Mars.

2.2 History of nuclear thermal propulsion

The concept of nuclear thermal propulsion was first publicly published by the Applied Physics Laboratory in 1947 [11]. Development of NTP systems began at Los Alamos Scientific Laboratory (LASL) in 1955 as Project Rover, under the auspices of the Atomic Energy Commission (AEC). NASA was formed in 1958 in response to Russia's launch of Sputnik and the beginning of the space race, and took over the Rover project with continued collaboration with LASL and the AEC [12]. Rover later became a civilian project within NASA and was reorganized to perform research directed toward producing a nuclear powered upper stage for the Saturn V rocket. In 1961, the NERVA program was formed by NASA to develop a nuclear thermal rocket engine. The program designed, assembled, and tested 20 nuclear rocket engines through a number of experimental series, including the KIWI, PEWEE, PHOEBUS, TF, and NRX reactors. These ground-based test reactors used solid fuel, based on advanced graphite materials, and were thermal spectrum reactors. The NRX-XE rocket reactor performed 28 burns with more than 3.5 h of operation [6], demonstrating the ability to operate and restart with the high performance requirements needed for use in an NTP system.

A Nerva-type engine concept is depicted in **Figure 1**. The fuel is manufactured as solid hexagonal blocks, with holes drilled through for hydrogen flow to cool the core. Multiple elements are assembled to create the core, with criticality control through the use of control drums with a poison plate on one side of the cylindrical drum, much as has been used at the Advanced Test Reactor (ATR) for over 50 years [14]. Minimal excess reactivity is needed as the total core lifetime will be on the order of hours, and will only operate for times on the order of an hour or less

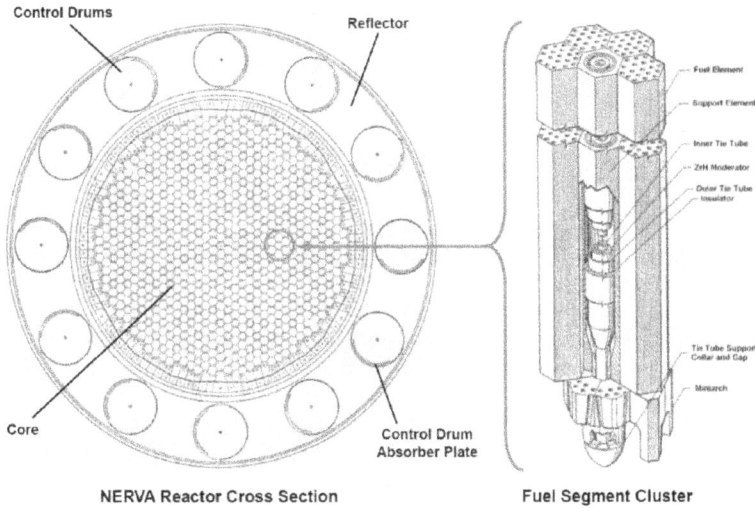

Figure 1.
Reactor core cross section for a ROVER-type NTP engine (left) and a cutaway of a fuel assembly cluster (right) [13].

resulting in minimal xenon buildup. These reactors were fueled using high-enriched uranium (HEU) in excess of 90% ^{235}U.

Both the Rover and NERVA research focused on a fuel form consisting of a graphite matrix with dispersed fuel (GMWDF). Graphite fuel compacts were used with various fuel types, including UO_2 and UC_2 fuel particles, and as $(U,Zr)C^1$ graphite composite. The three fuel forms used with the GMWDF compact are [15]:

- *Particles in graphite matrix:* This is the earliest compact form. It first contained UO_2 particles that were later replaced by UC_2 particles. This compact did not retain fission products and was soon abandoned.

- *Pyrolytic carbon (PyC):* PyC coated particles in a graphite matrix are the second generation of fuel used in the Rover and NERVA programs. This compact used UC_2 fuel particles and retained fission products well, but it features an inferior structural integrity as compared to the (U,Zr)C composites.

- *(U,Zr)C composite:* This is the most advanced of the GMWDF compacts with good structural integrity, closely matching thermal expansion coefficients between the composite and ZrC coating, as well as additional protection against corrosion by the carbide composite layers.

GMWDF compacts lead to a hard thermal spectrum [16–18]. Early designs exclusively used the graphite matrix as a moderator, but later designs starting with the PEWEE 1 experiment included ZrH sleeves in tie rods to increase the moderation ratio and reduce the core size [17]. The main issue with GMWDF compacts is that hot hydrogen corrodes the graphite matrix if they come into direct contact [15]. Therefore, all GMWDF compacts used coatings to protect the graphite matrix. The coatings must match the thermal expansion of the matrix closely to avoid excessive cracking. While still remaining a concern at the conclusion of project NERVA, corrosion rates were reduced by more than a factor of 10 [17]. GMWDF was used in

---

^{1} This notation denotes a solid solution where C sits on one lattice and U and Zr share the second lattice.

the shape of fuel plates (KIWI-A) and cylindrical fuel elements in a graphite module (e.g., KIWI-A' and KIWI A3), but most often as hexagonal fuel elements connected via tie tubes [17], as illustrated in **Figure 1**.

While not used in most early testing, CERamic-METallic (CERMET) fuels were evaluated during the NERVA program. The technology was too new and not well understood in the early 1960s, but was being investigated in parallel to the NERVA experiments. CERMET compacts consist of ceramic fuel particles embedded in a refractory metal matrix [19, 20]. The choice of matrix and fuel material influences thermal stability, thermal conductivity, structural integrity, and neutronics performance of the CERMET compact. Concurrent with the NERVA program, ANL and General Electrics (GE) developed separate CERMET NTP concepts. In a simplistic sense, CERMET fuels are particles of ceramic fuel (i.e., UO_2 or UN) encapsulated in a metal matrix, typically, but not limited to, tungsten, rhenium, or molybdenum. The research conducted by ANL and GE included the development and testing of the CERMET fuel and the design of the ANL-200, ANL-2000 [20, 21], and the GE 710 reactors [21, 23]. These CERMET programs focused entirely on HEU fuel kernels and fast reactor concepts. In contrast to GMWDF, the GE CERMET concepts did not undergo prototypical irradiation testing, nor did either concept undergo engine testing. Therefore, prior to the twenty-first century, the technology readiness of CERMET compacts trailed that of the GMWDF compacts.

The matrix material of a CERMET usually makes up about 30–60% of the compact volume [23], so its properties are both neutronically and structurally important. The ANL and GE programs focused mostly on natural tungsten as matrix material [20]. Among the available matrix materials, tungsten provides the largest fracture strength and temperature stability [6]. However, tungsten is brittle at low temperatures, causing issues with cracking. All isotopes of tungsten have strong (n, γ) resonances between 1 eV and 5 keV, thereby making tungsten neutronically challenging, except for fast reactor applications.

Fuel kernels also make up a significant fraction of the volume, so the materials properties and performance must be evaluated. Some work was performed in this area under the GE and ANL engine design programs for UO_2 and UN fuel types, as described below:

- *UO_2:* UO_2 fuel kernels were the only fuel form used in the ANL program and the primary fuel type pursued in the GE 710 project [15]. UO_2 has a uranium density of 9.7 g/cm^3. Both the ANL [20] and GE 710 programs used HEU enriched to 93%. For HEU CERMET compacts, the uranium content of the ceramic phase is not as important because enrichment can be adjusted to provide sufficient fissile material. The thermal conductivity of UO_2 is about 10 W/mK at room temperature and reduces with increasing temperature and burnup [24].

- UN: UN fuel kernels were considered as part of the GE 710 project [15]. UN has a uranium density of 10.7 g/cm^3. The thermal conductivity of UN is larger than that of UO_2 starting with a thermal conductivity of roughly 14 W/mK and increasing with temperature [25].

The GE experiments were at temperatures significantly lower than the NTP requirements, but provided much data on materials behavior and failure mechanisms [20, 26]. ANL focused on the production of CERMET fuels; different fuel fabrication procedures were employed with mixed success. Non-nuclear testing of samples was performed in flow loops of hydrogen heated to 2770°C to understand the fuel loss rates. Nuclear tests on the ANL CERMET samples were run in the Transient Reactor Test Facility (TREAT) located at Idaho National Laboratory

(INL). Eight specimen CERMET fuels, each with seven coolant holes, were tested under pulsed reactor conditions. Some fuel failure was observed in a few of the experiments [20, 26, 27].

With the success of the missions to the moon and the space race won after putting a man on the moon, the U.S. changed priorities for space exploration. Along with the cancelation of the Apollo missions, the NERVA program was terminated in 1972. Nevertheless, these programs provided a wealth of experience and knowledge; this work has been recently resurrected. Although the basics of rocket science have not changed since the 1970s, our understanding of materials performance and the development of new fabrication processes have advanced.

2.3 Current testing for NTP materials

Although historical experience in NTP design has provided a wealth of valuable data, recent advances in materials research have somewhat altered approaches to the design of NTP fuel, especially with respect to fuel material compositions, fabrication, and testing. Programs described earlier used HEU; current design concepts are based on high assay low enrichment uranium (HALEU) with a ^{235}U enrichment of less than 20% (often also referred to as simply LEU). Working with LEU greatly reduces security concerns and allows existing NASA facilities to work with fuel samples with minor modifications to address radiological concerns. HALEU would also be available at a significantly lower cost than that of HEU, and is much easier to transport. At the time of this writing, NASA is working with existing feed stock for test specimens as the U.S. cannot currently produce HALEU fuel. However, in June 2021, the U.S. Nuclear Regulatory Commission (NRC) approved a request from Centrus Energy to produce HALEU fuel at its enrichment facility in Piketon, Ohio. Centrus has already built 16 centrifuge machines for uranium enrichment, expecting to begin HALEU production by early 2022 [28]. This fuel will be used by both NASA and a number of advanced reactor prototypes under development in the U.S.

Current research and development efforts are organized within NASA's Space Technology Mission Directorate and are focused on both fabrication and performance under prototypical conditions. Although no NTP engine prototypes have been developed since the earlier work in Rover and NERVA programs, other facilities have been used for materials testing under reactor conditions. In early 2015, the first partial-length fuel elements were tested in the Nuclear Thermal Rocket Element Environmental Simulator (NTREES) located at NASA's Marshall Space Flight Center (MSFC) [29]. NTREES has been designed to provide up to 1.2 MW of heating to simulate an NTP thermal environment by capturing exposure to hydrogen heated to temperatures up to more than 3000 K. Numerous tests have been completed in NTREES; however, the facility is non-nuclear and unable to produce the intense neutron and gamma fluxes that will be present in an NTP engine. To that end, a number of tests have been completed or are planned for high-power transient tests in TREAT. In June 2019, the experiment designated as SIRIUS-CAL was the first test of an NTP-type fuel specimen. As with NTREES, a number of tests with representative fuel specimens have been completed and are ongoing.

To date, tests have been performed using CERMET fuel specimens based on fabrication experience gained in earlier ANL and GE CERMET tests, along with other facility tests. About 200 CERMET samples were tested in the various programs by thermally cycling to high temperatures in hydrogen [6], providing valuable data for performance and fabrication. CERMET fuel also allows for considerable control in fabrication due to the unique structure of the material itself.

Building on the earlier experience with natural tungsten as a matrix material, new materials have been evaluated:

- *Enriched tungsten:* Identical to standard tungsten, except that it is enriched in ^{184}W. While all isotopes of tungsten have strong (n, γ) resonances, ^{184}W has the least pronounced resonances with cross-sections smaller than 1000 barns and confined to energies between 10 and 500 eV.

- *Rhenium alloyed tungsten:* Rhenium increases the compact's ductility [15], but may reduce temperature limits [15].

- *Molybdenum:* Molybdenum compacts are more ductile than tungsten compacts, but are less heat resistant (e.g., they have a higher vapor pressure than tungsten) and prone to significant swelling induced by fission gas release [30]. Molybdenum has strong (n, γ) resonances between 10 eV and 50 keV that make it as neutronically challenging as tungsten.

- *Molybdenum-30 wt% tungsten:* Mo-30 W is of interest for moderated, LEU NTP reactors [31]. Mo-30 W is a good compromise between tungsten and molybdenum because its density is smaller, while its durability is just slightly below pure tungsten.

CERMETs can be used in LEU designs as discussed in Refs. [23, 32]. However, parasitic absorption of tungsten and ^{238}U, as well as reduction in fissile content, make it impossible to build a CERMET-based fast spectrum LEU core using natural tungsten [33]. Thus, the current focus of CERMET LEU cores is on thermal spectrum systems [23, 32]. For thermal reactors, CERMET offers the advantage of a higher fuel density as compared with composite GMWDF. However, the neutronic behavior of a CERMET compact is challenging because of parasitic absorption, the lack of moderating ability, and a short mean free path for the thermal neutrons [23]. These challenges result in difficulty adding reactivity to the core, requiring large fuel loadings and effective reflectors. They also exhibit significant self-shielding across fuel elements (with NERVA dimensions), leading to intra-element peaking and non-uniform burnup distributions after several tens of hours of operation [23].

NASA has been pursuing a parallel path in evaluation of CERMET- and CERCER (CERamic–CERamic)-based fuel forms. In 2021, NASA decided to place more emphasis on CERCER-based fuel concepts moving forward, although a number of CERMET-based fuel experiments are in the testing pipeline for the next few years. As opposed to CERMET, CERCER fuel requires approximately seven times less HALEU, has lower maximum fuel meat stresses, and is lighter [34]. CERCER fuels with coated fuel particles also offer the potential for increased margins with respect to fuel matrix melting compared to CERMET systems, but are at a lower level of technological and fabrication maturity. CERMET fabrication and testing began in the 1960s and 1970s for NTP applications, while CERCER (in NTP applications) is a relative newcomer [35]. The fabrication processes of CERCER fuels is currently based on relatively simple compression and sinter methods.

Both CERMET and CERCER fuels are being tested at both the TREAT and NTREES facilities. **Figure 2** illustrates the current plan for the experiments at TREAT with both CERCER and CERMET for the next several years. The CERMET tests have served as a technology pathfinder for CERCER fabrication and testing methods. The figure also shows the current plan for the testing program at TREAT, with experimental configurations becoming more complex, as well as plans to migrate from CERMET to CERCER fuels.

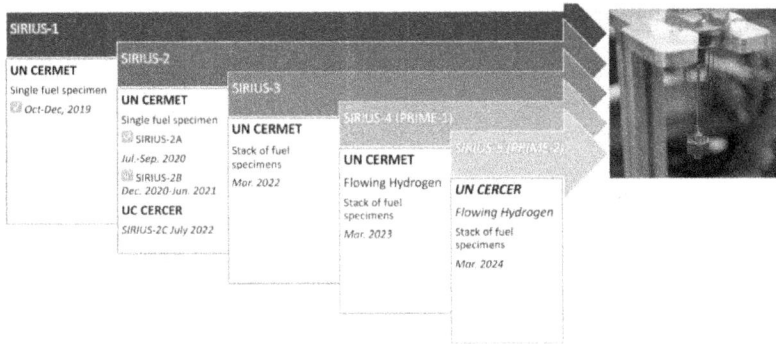

Figure 2.
Current high power neutronic testing plans; picture on the right shows the SIRIUS-1 test specimen being prepared for irradiation testing.

2.4 Needs for nuclear thermal propulsion material testing

The tests described in the previous section are being performed to collect information on the performance limits of fuel forms and cooling configurations. To meet mission requirements, it is desirable to maximize fuel temperatures, but higher temperatures introduce other issues: expansion, stresses, Doppler broadening, and chemical interactions. For the latter, early graphite fuel experiments under Rover highlighted the need to use coatings on the fuel grains. It is also known that fuel hydration from direct contact between fuel and hydrogen coolant has a deleterious effect on fuel performance [36]. Test specimens often include cladding materials on flow surfaces, which requires an additional evaluation in terms of clad/fuel interactions. Cladding is also an additional challenge in fuel fabrication. Cycling of the fuel from zero power to high power, operation at steady state for tens of minutes, and the return to zero power results in the potential buildup of temperature-driven stresses, which could ultimately lead to failure. Hence, material testing must address all of these physics, either in integral or separate effects testing. Both TREAT and NTREES provide capabilities for such tests. NTREES allows for larger specimen sizes and (until the SIRIUS-4 experiment is fabricated) is the only facility that provides for high temperature hydrogen flow. TREAT allows for direct nuclear testing with energy distributions that would be more typical of an NTP configuration. However, hydrogen flow within fuel specimens will be introduced within TREAT with the first Prototypic Reactor Irradiation for Multicomponent Examination (PRIME) experiments. PRIME-1 (also known as SIRIUS-4) will use CERMET fuel, while PRIME-2 will repeat the experiment with a CERCER fuel sample. Both are shown on the timeline in **Figure 2**. After PRIME-2, further experiments will focus on the evaluation of CERCER fuel specimens.

3. Overview of reactor types for NTP

A plethora of different NTP reactors were proposed and some of them were tested. Before considering particular examples, distinguishing features of reactor types are discussed. This allows for the development of a taxonomy of NTP reactors where one can more easily appreciate the differences in reactor physics characteristics and performance. We discuss different neutron spectra, fuel element geometry concepts, the use of low enriched and highly enriched uranium, fuel compact type, and the interplay of these factors when considering example designs.

3.1 Impact of the neutron spectrum

The advantage of thermal spectrum reactors is that criticality can be achieved with less fissile material in the core. In turn, the advantages of fast reactors are that no moderator is necessary, thereby allowing more space for fuel, and that the fuel matrix can be constructed from refractory metals without suffering from parasitic absorption at small neutron energies. Fast reactor designs are simpler and more robust because there is no need for a moderator that is either sensitive to elevated temperature, hot hydrogen, or both. In addition, the technological challenges of startup are smaller for fast reactors because of the smaller temperature defect and H_2 worth [37]. Finally, flooding with water leads to negative feedback effect in fast reactors [22].

Fast spectrum NTPs during the ANL-200/2000 and GE-710 projects were designed using HEU CERMET in hexagonal assemblies. It is impossible to achieve criticality in a fast reactor with LEU CERMETs [33]. However, it is possible to design a core with sufficient excess reactivity using UN fuel plates with refractory metal cladding [33]. This is enabled by the much smaller ratio of refractory metal to fuel volume than in the LEU CERMETs.

3.2 Neutronics parameters of interest

The moderator-to-fuel density ratio (MTFR) [38] is an important characteristic for the reactivity of a reactor. There exists an MTFR at which the core multiplication factor assumes a maximum and the core is optimally moderated, while for smaller or larger MTFRs, the core is undermoderated or overmoderated, respectively. From a control perspective, it would be desirable to have an undermoderated core to avoid positive feedback from increasing power. For overmoderated reactors, reduction in hydrogen density caused by an increase in power can lead to a positive reactivity feedback loop. NERVA and derived designs are all undermoderated, as the addition of hydrogen leads to an increase in core reactivity [39]. For LEU reactors, multiplication factor, size, weight, and thermodynamic performance depend heavily on the moderator-to-fuel ratio [40].

Power peaking measures how uniformly the power is produced in the core, and can be computed by taking the maximum power density observed in the reactor and dividing it by the average power density [41]. In practice, it is more common to consider fuel element or fuel assembly peaking, and considering both axial and radial components. These are computed by taking the maximum fuel element power and dividing it by the average fuel element power. The importance of the power peaking is that limiting core conditions, such as peak temperatures, are usually experienced in peak fuel elements.

The temperature peaking factor is related to the power peaking factor, but is influenced by both the power peaking and thermal-fluid conditions in the core. It is defined as the peak fuel element temperature divided by the average temperature of the fuel compacts. Larger power peaking factors can be addressed by directing more flow to the high-power regions, which leads to reduced temperature peaking factors.

Reactivity feedback is the effect that non-neutronic parameters have on the reactivity of the core. When reactivity is positive, reactor power increases, while the opposite is true for negative reactivity. The most important feedback mechanism and the parameters to which they are sensitive are:

- *Doppler Broadening:* Doppler broadening increases the absorption by increasing resonance width with increasing material temperature [38]. While any material with absorption resonances exhibits Doppler broadening, the most prominent effect is usually stemming from ^{238}U. Due to the much larger amount of ^{238}U relative to HEU, Doppler broadening is much more important in LEU reactor concepts. Doppler broadening is always a negative feedback effect with increasing temperature and is effectively an immediate effect with respect to the temperature of the fuel.

• *Spectral Shift:* Spectral shift refers to the hardening of the spectrum when the moderator temperature is increased. Spectral shift has a negative feedback effect in graphite moderated HEU compacts [42]. However, spectral shift in reflectors can improve reflector efficiency and increase reactivity.

• *Thermal Expansion:* Thermal expansion of the core occurs due to an increase in the core temperature. Thermal expansion leads to an increase in surface area for the fuel in the core, increasing leakage at the expense of fission. Thermal expansion is sensitive to material temperatures in the core, the mechanical design of the core and the expansion coefficients of the materials. Thermal expansion is always a negative feedback effect.

• *Hydrogen Moderation:* Hydrogen in the core moderates neutrons and leads to a softer spectrum; hence, the probability of fission ^{235}U increases, while the likelihood of resonance absorption and leakage decreases. Therefore, increased hydrogen flow is usually a positive feedback effect. However, if the core is already past its optimal moderator-to-fuel ratio, the addition of more hydrogen leads to an increase in parasitic absorption in the moderator, and consequently, a reduction in reactivity.

• *Fuel Burnup:* Fuel burnup is the consumption of nuclear fuel and the production of direct and indirect fission products. It influences the reactivity by removing fissionable material and adding potential absorbers. The effect of burnup is very slow, on the order of the lifetime of the NTP system.

Fast reactors have the smallest feedback coefficients. Burnup and hydrogen content do not have an appreciable effect, while temperature via expansion and Doppler and spectral shift have a comparatively small and equal effect. HALEU-fueled reactors react predominantly to temperature via the Doppler/spectral shift. Burnup affects reactors with smaller loading of fissile isotopes more than reactors with higher fissile loading (e.g., GWDF typically has a smaller fuel loading than CERMET). The largest feedback effect for HEU GWDF is the hydrogen content of the core because Doppler broadening effects are small and the spectral shift is not as strong a feedback mechanism as that of hydrogen. Note that the sensitivity of the reactor to hydrogen content is used to introduce positive reactivity into the core by increasing the flow. The large positive reactivity coefficient does not make the HEU GWDF core dynamically unstable because an increase in reactor power leads to a reduction in hydrogen density, and thus, a negative feedback effect. Note that many observations here are based on feedback effects tabulated in Ref. [37].

Feedback is important for the controllability of the core. Large negative feedback coefficients as present in LEU cores with respect to fuel temperature require the control mechanism to have sufficient excess reactivity in reserve; therefore, thermal NTP reactors, especially if LEU fueled, must have control mechanisms with a much larger magnitude of reactivity relative to fast systems.

3.3 Geometrical arrangements

This section introduces different criteria to distinguish and classify geometries in NTP reactors. The criteria we use to distinguish these reactors are the fuel element geometry (e.g., hexagonal, annular, plates), the structural concept (tie tube or monolith), and if the moderator is heterogeneous or homogeneous. Here we compare U.S. NTP designs to concepts evaluated in the Soviet Union and the Republic of Korea. The Soviet Union began at about the same time as the Rover program, but ended in 1989 with the collapse of the Union of Soviet Socialist Republics (USSR). The Korean concept is still under active development, beginning in 2013.

NTP reactors are distinguished by their fuel element layout. The original NERVA design used hexagonal fuel elements arranged in a hexagonal lattice, as shown in **Figure 3(a)** and **(b)**. A group of six fuel elements is connected to a tie tube. The tie tube is relevant both for moderation and structural integrity as discussed below. The ratio of the number of fuel elements and tie tubes in the lattice is an important parameter for NERVA-type designs.

Fast reactor concepts originating from the ANL and GE projects also use hexagonal fuel elements, as observed in **Figure 3(c)**, arranged in a hexagonal lattice, as seen in **Figure 3(d)**, but the fuel elements tend to be larger than their NERVA counterparts and contain more coolant channels. The hexagonal fast concepts do not require tie tubes.

The Russian NTP program considered a variety of fuel element shapes (see Ref. [15]) among which the twisted ribbon design depicted in **Figure 3(e)** was selected as the most promising option. Usually, each twisted ribbon is referred to as a fuel element; it should be noted that each twisted ribbon is significantly smaller than a NERVA fuel assembly. Twisted ribbons are inserted into a fuel bundle that is wrapped by insulating material. The fuel bundle is in turn inserted into a fuel assembly that is then placed into the core.

The Korea Advanced NUclear Thermal Engine Rocket (KANUTER) fuel assembly design is depicted in **Figure 3(f)**. The fuel shown in red in the figure consists of wavers forming square flow channels; interlocking of the fuel wavers forms a square lattice [45]. The fuel is surrounded by insulating carbon wrappers and a metal hydride moderator. The fuel assemblies in the KANUTER core are arranged in a hexagonal pattern.

The recent NASA/BWXT design is depicted in **Figure 3(g)** with the progression from the smallest to largest part from left to right in the figure. Each fuel element is cylindrical with round flow channels and is surrounded by an insulator. The flow channels in each element are arranged in cylindrical clusters in CANDU reactors (i.e., one central hole and six flow channels placed on a circle around the center with 12 flow channels placed on a larger circle surrounding those, etc.). The fuel elements are wrapped with structural support and then placed in holes bored through the monolithic core structure, as observed in the second picture from the right in **Figure 3(g)**. The monolithic core structure is made up of a metal hydride moderator. The fuel elements in the monolith are arranged in a cylindrical cluster, just as the coolant channels are arranged in the fuel element.

The core geometry can be distinguished by the structural support concept for the fuel elements. In the NERVA designs, a tie tube is connected to the six fuel assemblies around it, and a spring keeps the fuel elements in tension to avoid damage to the core structure by flow-induced vibrations and support the core against the axial pressure drop [47]. The tie tubes are connected to a support plate located at the cold end of the core. Additional axial support is provided by pedestals in some reactors (e.g., PEWEE) [17].

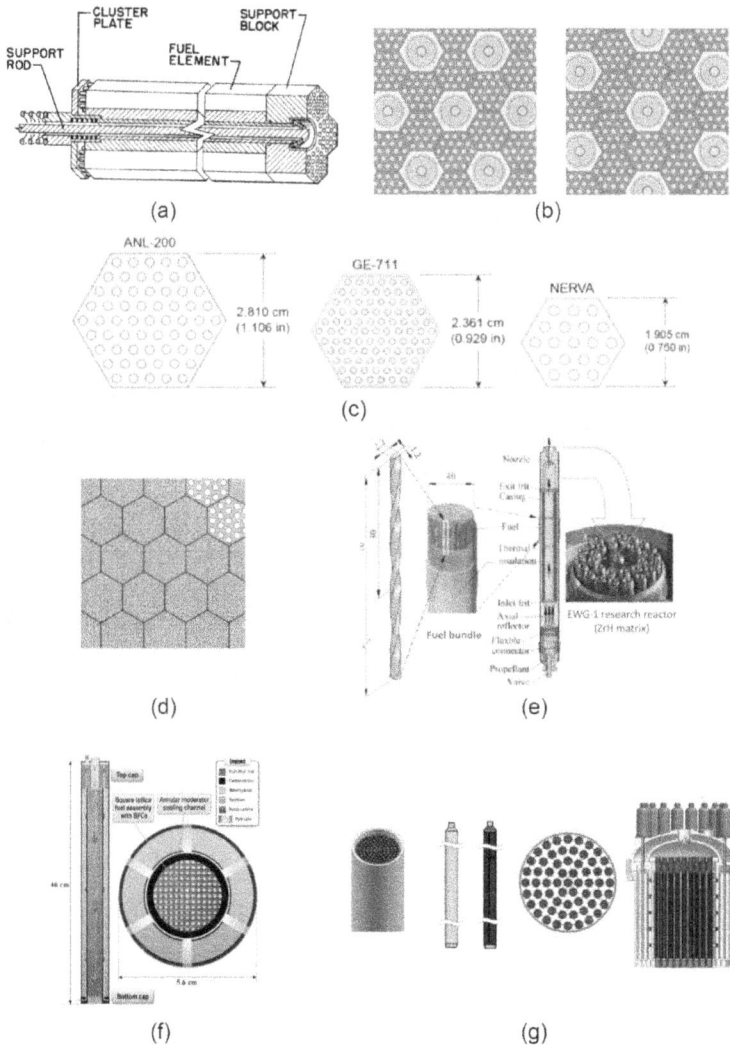

(a)

(b)

(c)

(d)

(e)

(f)

(g)

Figure 3.
An overview of the geometric arrangement of different NTP concepts. (a) Typical later NERVA fuel element layout. Six fuel elements are connected to a tie-tube [16]. (b) NERVA hexagonal fuel element layout with a different ratio of tie-tube and fuel elements [43]. (c) ANL-200, GE-711, and NERVA fuel assembly geometries [44]. (d) Hexagonal lattice of fuel elements typical for fast reactor designs like ANL-200 and GE-710 [32]. (e) Russian NTP concepts using a twisted ribbon fuel element in an encased assembly that is inserted into the reactor (picture (e) from left to right) [39] (length units are mm). (f) Korean integrated fuel assembly design with square flow channels [45]. (g) Recent "fuel assemblies under consideration for NASA's nuclear thermal propulsion reactor designs" by BWXT advanced technologies, LLC [46].

In contrast to the tie tube design, the more recent NASA/BWXT design uses a monolith concept, as described in Ref. [46] and shown in **Figure 3(d)**. The monolithic core structure is made up of the metal hydride moderator and has borings that contain the fuel assemblies. The fuel elements are wrapped with insulator and structural support. The structural support is fastened to a support plate at the cold end of the core. Additional axial support at the cold end may be included in the design as well.

Finally, the KANUTER design, as shown in **Figure 3(f)**, arranges beryllium spacers between the integrated fuel assemblies. In contrast to the NASA/BWXT design, the integrated fuel assemblies contain a moderator where the core support structure is strongly moderating.

Reactors can also be classified by how the moderator and fuel are arranged. If the fuel and moderator are spatially separated, the reactor is *heterogeneous*; if the fuel and moderator are mixed, then the reactor is *homogeneous*. For this distinction, spatially separated means that there is sufficient distance on the order of a mean free path between the fuel and the moderator. Heterogeneous cores offer an advantage in reactivity over spatially homogeneous cores; the effect is sometimes referred to as fuel lumping. If the moderator is spatially separated from the fuel, then moderation happens away from the fuel, reducing the likelihood of resonance absorption during the slowing down process [48].

To the knowledge of the authors, the only truly *homogeneous* cores were early NERVA designs before PEWEE 1. In these designs, the moderator was the graphite matrix containing the fuel particles. Starting from PEWEE 1, the tie tubes were equipped with ZrH sleeves adding additional moderation to the system and making these designs essentially *mixed* moderation cores [17]. The Russian cores and KANUTER are *mixed* moderation cores due to the presence of graphite in the fuel compact (i.e., the homogeneous portion) and an additional moderator either in the fuel assembly or the structural components surrounding them. The recent NASA/BWXT design is a *heterogeneous* core because the only significant amount of moderator is in the monolith outside of the fuel assemblies. Fast reactors do not fall into this classification because they do not contain a moderator.

The following section discusses a small selection of representative NTP reactor concepts and provides more detail on each design.

3.4 Reactor concepts

PEWEE-1 is a demonstration reactor tested in the NERVA program in 1968 toward the end of the program. It is a small reactor when compared with the preceding Phoebus tests with power reduced from 4000 MW in the Phoebus-A design to about 500 MW [17]. To offset the increased leakage from the smaller core size, ZrH sleeves were inserted into the standard tie-tube concept of the NERVA program; the tie-tube ratio (TTR)^2 was increased and the reflector thickness was increased. The main objective of PEWEE-1 was to serve as a test bed for fuel elements and no attempt was made to maximize the outlet temperature [17]. Despite these differences to other tests within the NERVA project, PEWEE-1 is a good example of the technology used and resulting observed performance during NERVA.

In two different works [49, 50], Kotlyar focuses on studying the design space of thermal LEU-CERMET NTP concepts. These designs use the NERVA structural concept of fuel elements and tie tube/moderating elements without changing their size and shape (i.e., a hexagonal lattice with 1.905 cm flat-to-flat distance). However, the matrix is changed to LEU UO_2 particles in W-CERMET [49] and LEU UN particles in Mo or MoW-CERMET [50]. In order to overcome the reactivity penalty of refractory metals, lower uranium enrichment, and the lack of moderation in the fuel compact, Kotlyar's core concepts include significantly more moderating elements (>50% depending on core size) than PEWEE-1 with more ZrH moderator and additional carbon per moderating element. The spectrum is more thermal than in the NERVA engines, but is significantly undermoderated for the optimal small, medium, and large NTP designs [49].

---

^{2} The ratio of tie-tube elements to total number of elements.

KANUTER [45] is unique among modern NTP designs because it uses HEU with an enrichment of 93%. The goals of the design are to maximize I_{sp}, thrust-to-weight ratio, and allow for bimodal operation (e.g., thrust and electricity generation). The NTP design uses a tricarbide (U,Zr,Nb)C fuel matrix that was tested during the Russian NTP program. KANUTER uses an integrated fuel assembly concept; the fuel assembly depicted in **Figure 3(f)** contains both fuel and a ZrH moderator that are separated by carbon–carbon insulation. The fuel matrix is arranged in wafers and the coolant channels are square. In the core, the 37 fuel elements are arranged in a hexagonal lattice and held in place by cooled beryllium spaces.

Poston [32] investigated how the performance characteristics of NTP systems change when the fuel matrix is changed from GWDF to CERMET and the enrichment is changed from LEU (19%) to HEU (93%). The four variants discussed in Ref. [32] are thus HEU-composite (e.g., NERVA carbide composite fuel), LEU-composite, LEU-CERMET, and HEU-CERMET. All concepts use hexagonal assemblies, but the assembly sizes differ: the HEU-composite uses the standard NERVA 19-hole element with a 1.91 cm flat-to-flat, the LEU-composite uses a 37-hole fuel element with a 2.77 cm flat-to-flat, the HEU-CERMET uses an element similar to the GE-710 designs with 91 holes and a 2.57 cm flat-to-flat, and the LEU-CERMET uses a 61 hole assembly with a 2.52 cm flat-to-flat. With the exception of the HEU-CERMET, all designs use the traditional fuel element/tie-tube concept of NASA albeit at different TTR (33% for HEU and LEU composite and 50% for LEU-CERMET). All concepts have an epithermal spectrum except for the HEU-CERMET. Moderation in the epithermal concepts is provided by the composite and by ZrH in the tie tubes; the LEU-CERMET requires more tie-tubes to increase the amount of moderator in the core. The CERMET in Poston's study is enriched to remove the highest absorbing isotopes from tungsten, molybdenum, rhenium, and zirconium; tungsten is used as a matrix material in the study. All designs use a Be radial reflector and the CERMET designs use a BeO top (cold-end) reflector. The performance difference and differences in the design parameters depend most heavily on ^{235}U densities. The neutronics design ensures a 1% beginning of life reactivity margin and a shutdown margin of 5%; however, LEU-CERMET barely achieves the beginning of life margin.

In Ref. [33], Youinou evaluates alternative designs to the monolithic ZrH moderated, CERMET, or CERCER concepts of the early 2020s by NASA. While several different concepts of this report deserve attention, the most important design is an LEU, plate-fueled, fast design. This concept uses UN fuel plates of thicknesses 0.5–10 mm, MoW or W clad of thickness 0.25–0.5 mm, square assemblies of size $8 \times 8 \times 80$ cm, and 7–49 fuel plates per assembly. There are 37 fuel elements in the core. The core has a power of 250 MW generating a thrust of 12,500 lbs. Youinou found that the smaller fraction of refractory metals in the plate design allow for fast LEU NTPs fueled with UN and clad with refractory metals.

The GE-710 NTP system is an example of an HEU, fast, CERMET-based concept that was developed concurrently with the graphite-based NERVA concepts [22]. The GE-710 program tested various CERMET matrix materials, including tungsten, tungsten-rhenium, tungsten-rhenium-molybdenum, and molybdenum-rhenium, among others [22]. All fuel elements investigated during the GE-710 are hexagonal and slightly larger than the NERVA fuel elements (e.g., 2.36 cm versus 1.91 cm flat-to-flat). GE-710 elements contain significantly more coolant channels than the NERVA elements, which increases the pressure drop through the core, but decreases the difference between the coolant and the maximum fuel temperature. Overall, the GE-710 project demonstrated excellent thermal and mechanical stability during thousands of hours of testing [51].

4. Modeling and simulation of NTPs

In this section, we focus on the modeling and simulation (M&S) needs for NTP systems from a nuclear reactor perspective, with a particular emphasis on transient modeling. INL leads the development of the multiphysics object oriented simulation environment (MOOSE) [52] that provides a cohesive framework for multiphysics analysis; MOOSE is introduced first. The needs of a transient reactor-centric M&S are introduced next, and then MOOSE applications performing transient simulations are introduced. Finally, we present the capabilities of MOOSE for a PID controlled startup transient.

4.1 Multiphysics object oriented simulation environment

MOOSE is a C++ based framework for a finite element and finite volume-based solution of partial differential equations. Its goal is to provide high-level access to the powerful finite element capabilities implemented in the libMesh library [53] and the linear and nonlinear solver technologies in PETSc [54] without having to understand multiple interfaces, manage parallel execution, or handle input/output. MOOSE is structured such that code can be reused among different research groups, facilitating the development of a multiphysics ecosystem referred to as the MOOSE herd.

The MOOSE framework provides: (1) extensible systems that perform tasks in a partial differential equation (PDE) solver and can be inherited from and used by physics applications; (2) an input/output handling system; and (3) specific internal data structures like the finite element mesh and finite element variables. Physics applications are developed on top of the framework. To date, the MOOSE repository comes with 21 modules (i.e., open-source physics implementations that are general enough to be packaged with MOOSE) including heat conduction, Navier–Stokes, and phase field. Many physics applications have been created based on MOOSE that contain either export-controlled, proprietary, or very specialized physics and require user approval and licensing.

The difference between MOOSE and traditional multiphysics nuclear engineering applications is that MOOSE is not a collection of single-physics codes connected with *glue code* [55]. MOOSE-based software applications are built using interfaces provided by the framework that are extended and specialized using inheritance. This paradigm shift away from using glue code provides many advantages, including reduction in data storage duplication, increased robustness against future compatibility issues, shared representation of geometry precludes developing a significant number of translation routines [56].

4.2 Relevant physics and simulation capability within MOOSE

4.2.1 Neutronics

Neutronics is at the heart of a reactor-centric viewpoint of NTP M&S. The neutron distribution drives the power distribution, which in turn drives temperatures and stresses in the core. In addition, the dynamic behavior of NTPs is to a significant degree driven by the neutronics feedback behavior. In contrast to most terrestrial reactors, NTPs spend a large fraction of their operating life in transient operation. Therefore, neutronics M&S for NTPs should provide a strong transient simulation capability. Traditionally, many neutronics tools are developed for steady-state (i.e., k-eigenmode calculations) or very slow transients (i.e., depletion calculations). During a transient, temperature and thermal-fluids conditions can vary rapidly, making a tight coupling of neutronics and heat conduction mandatory. Finally, one of the control mechanisms for the ramp up to power is the rotation of the control drums. During a startup transient, the neutronics code must be able to accurately model the behavior of the control drums rotated to an arbitrary angle.

Griffin is the MOOSE-based reactor multiphysics application [57]. It is a superset of the capabilities previously implemented in Rattlesnake [56] and Proteus [58]. In the near-term, it will also provide an interface to the MC2–3 cross-section preparation capability [59]. The main distinction between Griffin and traditional radiation transport solvers is that it is designed for transient multiphysics simulations, making it an ideal candidate for NTP simulation. Griffin is a deterministic radiation transport application that provides the user with a variety of solvers for the linear Boltzmann transport equation. These solvers provide a variety of different fidelity levels ranging from zero-dimensional point-kinetics models over neutron diffusion with equivalence correction to high-fidelity S_N models [56, 57] with spatial kinetics.

Griffin is an ideal candidate for transient analysis of NTPs. It naturally couples to MOOSE's heat conduction capability, described later in Section 4.2.2, and can be either connected via a Newton scheme (full coupling) or a Picard iterative scheme (tight coupling). It provides several radiation transport methods that can be used in steady-state and transient analysis with general cross-section and geometric feedback. For transient simulations, cross sections are usually pre-tabulated and then interpolated during the transient. Griffin provides a control-drum decusping method that allows an accurate modeling of control drum motion during a transient simulation [60].

4.2.2 Heat conduction and conjugate heat transfer

The temperature distribution in NTPs is of great importance. First, it is the most important driver for neutronics feedback in thermal LEU-fueled reactors, and second, temperature values and differences (cold to hot) are large and margins to failure are typically small. During normal operation, most heat is transferred to the hydrogen via conjugate heat transfer. However, some of the heat is transferred from the fuel through the insulator and multiple gas gaps to the moderator and even to the reflector. Heat transfer through the gas is mostly facilitated by radiation. In addition to heat transfer from the fuel, some of the fission heat is deposited non-locally in the moderator and reflector; it is therefore required to model a significant portion of the core to obtain an accurate understanding of the temperature distribution in the moderator and reflector.

Heat conduction in an NTP needs to consider the change in thermal properties with temperatures. The temperatures over the time of a startup transient and at different locations within the reactor vary significantly. The material properties relevant for thermal analysis of the problem (e.g., thermal conductivity, specific heat, density) vary as a function of the temperature, thereby requiring an accurate model that can account for the temperature dependence of these properties. NTPs use a significant number of special purpose materials (e.g., porous ZrC insulator, refractory metal matrix with uranium inclusions) and the thermal properties of these materials need to be available.

Heat transfer in open spaces of the reactor (e.g., plena and exhaust nozzle) must also model thermal radiation in complex geometries. MOOSE provides heat conduction, gap heat transfer, and net radiation transfer capabilities within its heat conduction module. The material system in MOOSE has the ability to use general temperature-dependent material properties supplied as polynomial fits, lookup tables, or customized material implemented in MOOSE source code. The BISON fuel performance code provides a variety of material models for nuclear materials [61]. BISON offers material properties for W and Mo-30W CERMETs [62].

The heat conduction module provides different interfaces for representing conjugate heat transfer. It can be applied as a boundary condition on channel boundaries or it can be lumped into a volumetric term. The coupling with the thermal-fluids code RELAP-7 [63] can be performed using a Robin-Neumann boundary or a Robin-Robin boundary strategy.

4.2.3 Thermal mechanics

Stresses in NTP systems arise from large temperature gradients, mechanical contact during transient and steady-state operation, and pressure differential over the core. The mechanical problem is a coupled problem between heat conduction, mechanics, contact, and potentially thermal-fluids. Vibrations can manifest in the solid structures that interact with fluid pressure oscillations caused by turbomachinery, flow separation, or other fluid-mechanical effects. The material properties relevant in mechanical problems include Young's modulus, Poisson's ratio, the linear expansion coefficient, and parameters describing plastic deformation, such as the yield stress and hardening law; these material properties generally depend on temperature.

MOOSE provides the capability to conduct mechanics simulations in the *tensor mechanics* module [64]. The tensor mechanics module is seamlessly able to couple with the heat conduction module, facilitating thermal-mechanics simulations. MOOSE also implements a variety of mechanical contact algorithms in its *contact* module. Finally, MOOSE allows pluggable multiphysics capabilities coupling neutronics, heat conduction, and time-dependent mechanics [65].

4.2.4 Thermal fluids and balance of plant

Nuclear thermal propulsion in its current form in the U.S. uses a HALEU-based reactor core to generate several hundred megajoules of thermal energy to heat hydrogen propellant to high exhaust temperatures for engine thrust. NERVA designs up to current engine concepts are of an *expander cycle* design; **Figure 4** shows a simplified representation of an NTP expander cycle engine.

In this design, high pressure liquid hydrogen (H_2) is pumped from storage tanks and is preheated while used to cool the nozzle, reactor pressure vessel, reflector and control drums and control drums (converting it to gaseous H_2), using the energy added to the gas to drive turbines. The exhaust from the turbine is directed to core support and shielding structures (not shown in **Figure 4**). Next, the gas passes through the coolant channels in the individual coolant block comprising the reactor core, where it is superheated to the necessary high exhaust temperatures. Finally, the gas is expanded through a nozzle with a high nozzle area ratio to generate thrust. Thrust is maximized by maximizing the gas temperature exiting the core, but current reactor material performance limits will restrict the peak temperature to something less than about 3000 K [44].

Unlike power reactors, NTP engines are expected to operate continuously for less than an hour at a time with weeks to months between burns [66]. Each operational period will consist of three phases: startup to full power, full thrust operation, and shutdown (with decay heat removal). Flow rates are matched to the reactor power according to the demands of each period. During startup, hydrogen economy requires as rapid an ascent to full power as possible through appropriate control drum rotation, and H_2 flow is used to both cool the reactor, as well as protect other

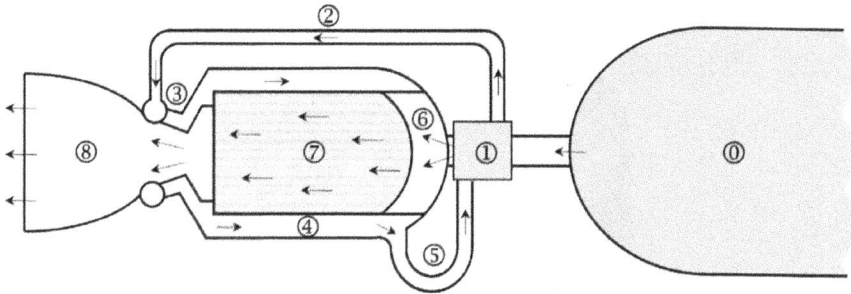

Figure 4.
Representation of NTP engine system with (0) liquid hydrogen storage tank, (1) pre-heated-hydrogen-driven turbopump, (3) nozzle cooling, (4) pressure vessel/reflector/control drum cooling, (5) gaseous hydrogen feed to turbopump, (6) gas plenum above core, (7) reactor core and hydrogen cooling, and (8) exhaust nozzle.

engine components. During the full thrust period, the core and balance of plant are near steady-state conditions. At shutdown, the reactor will be returned to a sub-critical state, but hydrogen flow will be needed for decay heat removal.

The M&S capabilities required for the thermal-fluids and balance of plant are: ability to exchange heat with solid conduction (i.e., conjugate heat transfer), modeling hydrogen in a temperature range from 40 to >3000 (or greater than 3000) K, ability to model compressible flow, availability or extendability to include heat transfer and pressure drop correlations suitable for NTPs, ability to model the relevant components in the NTP system (e.g., turbo pump and turbine on common shaft, valves, etc.), and a flexible control system that allows for the simulation of complex controllers.

RELAP-7 can solve single-phase (e.g., 3-equation model) and two-phase (e.g., 7-equation model) system analysis problems using a discontinuous Galerkin HLLC (Harten, Lax, and Van Leer Contact) discretization [67]. RELAP-7 provides models for a variety of components, including pipes, pumps, valves, and turbines; in addition, it supports both full (i.e., single nonlinear problem) and tight (i.e., Picard type) coupling with MOOSE heat conduction solvers via conjugate heat transfer. RELAP-7 provides *para-hydrogen* fluid properties across the required range, and provides a flexible and extendable control system that can be used to simulate the control system for an NTP model.

4.3 Case study of a reactor startup simulation with MOOSE

In this section, Griffin, RELAP-7, and MOOSE modules are coupled and used for a simulated startup of a LEU, CERMET-based core similar to the one depicted in **Figure 5**, but with an operating power of 250 MW and an approximate thrust of 55,600 N (12,500 lbf). The core consists of 61 LEU fuel assemblies arranged in five circular rings within a zirconium hydride (ZrH) monolithic moderator block. The startup simulation includes a PID-controlled rotation of the drums to match a predetermined reactivity setpoint curve, neutronics modeled with diffusion and Super-Homogenization (SPH) [68], heat conduction, and thermal-fluids.

From a neutronics standpoint, the probabilities of neutron interaction represented by cross-sections are affected by several temperature-driven feedback mechanisms. For the reactor shown in **Figure 5**, the primary feedback comes from the increase in ^{238}U capture reactions as the fuel heats up (Doppler feedback) while other important feedback mechanisms are spectral shift and hydrogen content in the core. From a modeling perspective, spectral shift and changes in moderator content are more difficult, because their effect is global. The value of the

Figure 5.
Concept of BWXT NTP reactor design (picture courtesy [34]).

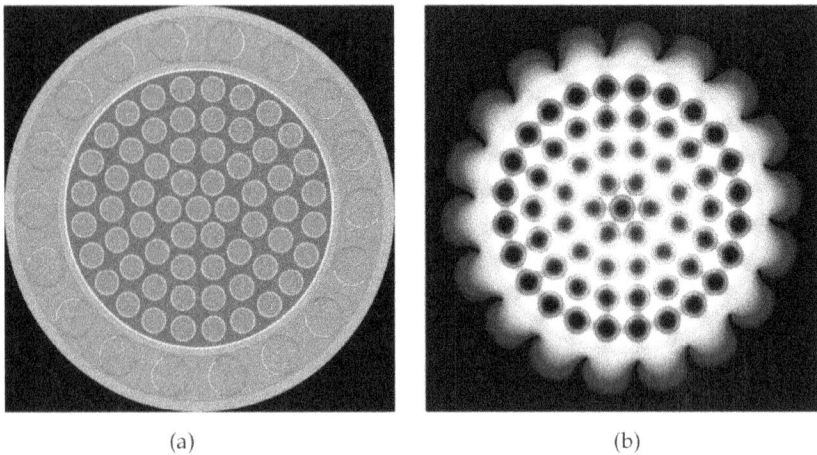

(a)　　　　　　　　　　　　　　　　(b)

Figure 6.
Full-core serpent model. (a) Geometry; (b) fission rate and thermal flux.

temperature or hydrogen density at one point affects the neutron spectrum, and thus, the effective cross sections at another point. The distance over which non-local effects materialize depend on how far a neutron can travel without being absorbed or escape the reactor. For this kind of reactor, this travel distance, or mean free path, can be quite large (on the order of 1–100 cm, depending on the neutron energy) and complicates the cross-section evaluation significantly, especially as the tremendous axial thermal gradient gives perceptibly different neutron spectra in different parts of the core. Therefore, the analyst may opt to tabulate cross sections not only for different temperatures and hydrogen, but also for different shapes of the temperatures and hydrogen densities. For this example, cross sections are pre-tabulated for different values of the important feedback variables (e.g., fuel, moderator and reflector temperatures, control drum angle). The Serpent Monte Carlo code is used for tabulating the cross-sections for this work [69] Plots from the Serpent model are shown in **Figure 6**.

The accuracy of the solution and execution time of the model are balanced by representing the neutron distribution by the neutron diffusion equation, discretizing it on a coarse mesh, and using the full-core SPH in Griffin. SPH can be seen as a physics-based reduced order modeling approach. This enables the use of a coarse numerical mesh, as shown in **Figure 7**, while preserving the key quantities of interest needed for the multiphysics coupling, such as reactivity and power density distribution.

The moderator monolith is not expected to see a large temperature increase compared with the fuel because each of the fuel assembly is surrounded by a layer of insulator. For preliminary calculations, it is thus acceptable to assume that fuel assemblies exchange little heat with one another. Due to various symmetries, the conductive and radiative heat transfer over each ring of fuel assemblies is therefore simulated by a single 30° slice, shown in **Figure 8** and extruded over the entire height of the active core. In this figure, the orange, red, green and blue regions correspond to the fuel, insulator (ZrC), shell (SiC), and moderator (ZrH), respectively. The fuel region is penetrated by 127 cooling channels. The moderator is also cooled by flow channels to remove most of the heat that radiatively crosses the three gaps between the fuel and the moderator. The thermal-fluids is modeled by two representative cooling channels per fuel assembly ring to simulate the convective heat removal in the fuel and in the moderator.

The integration of the various sub-modules into a multiphysics model is summarized in **Figure 9**. The neutronics model provides the power density into each of the 30° slice thermal models (e.g., one per ring). These provide the wall temperature to their respective cooling channels, which in turn provide the fluid temperature and heat transfer coefficient needed to evaluate the amount of heat removed by the coolant. Once the thermal field in each of the representative fuel assemblies is obtained, the fuel and moderator temperatures are passed back to the neutronics model to update the cross sections accordingly.

To perform a reactor start-up, the control drums need to be rotated to add sufficient reactivity to not only increase the reactor power, but also compensate for the negative feedback ensuing from the heat-up of the fuel. Attempting to select the rotation of the drums *a priori* to obtain a desired power evolution would likely require significant trial and error iterations, especially considering the nonlinear behavior of the reactivity feedback coefficients and fuel heat capacity as a function of temperature. Rather, efficient control of the drums can be achieved through automated means—for instance relying on a widely-used Proportional-Integral-Derivative (PID) controller, as illustrated in **Figure 10**. Given a desired power set-point, it can be converted into a reactivity signal ($\tilde{\rho}$ in **Figure 10**), which is then compared to the *measured* reactivity from the model. This measurement

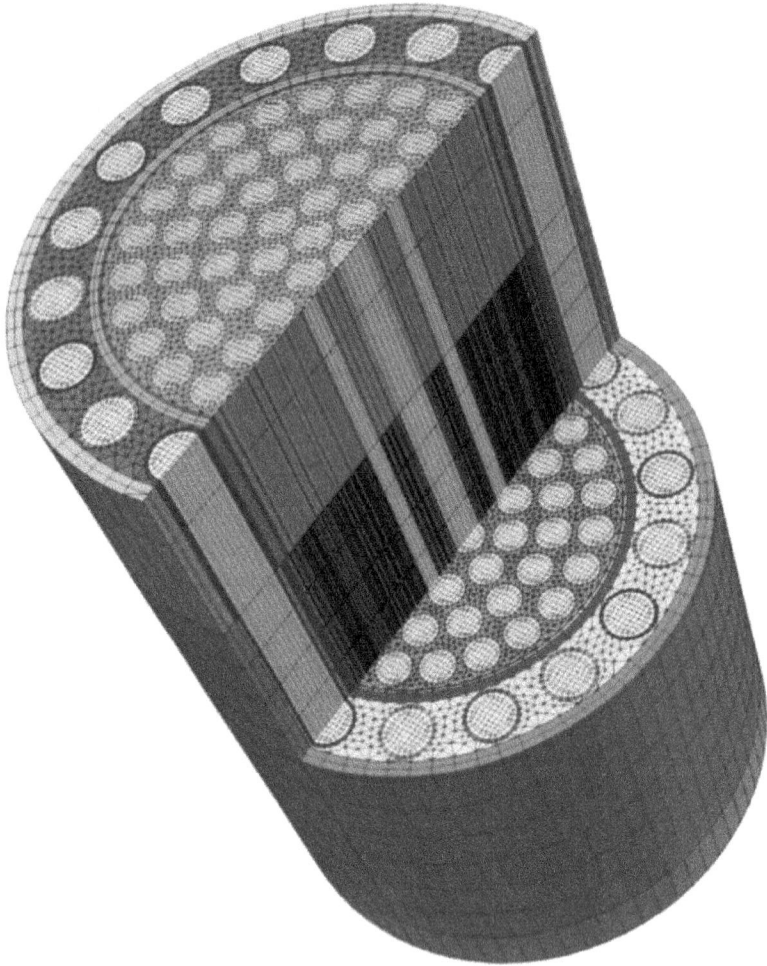

Figure 7.
Full-core neutronics and thermal meshes.

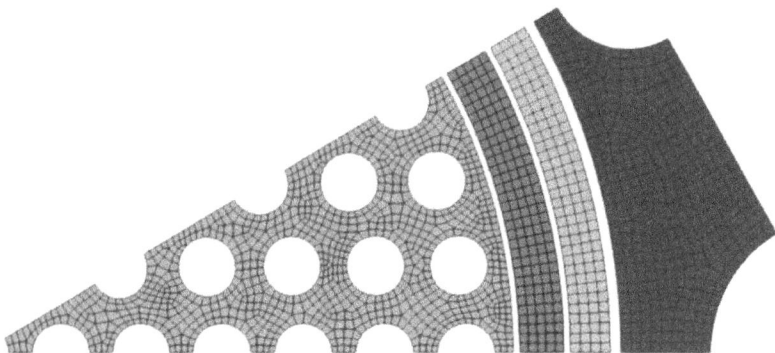

Figure 8.
X-Y view of the 30° slice thermal mesh.

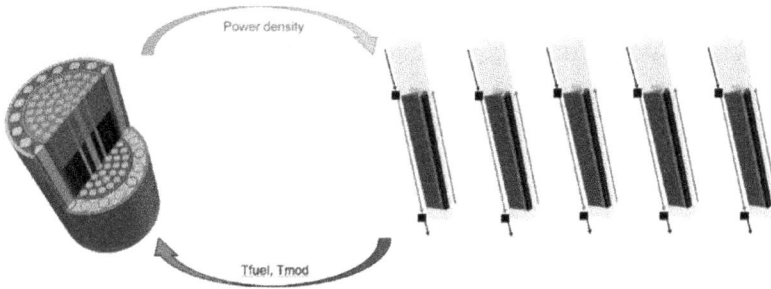

Figure 9.
Schematics of the full-core multiphysics model.

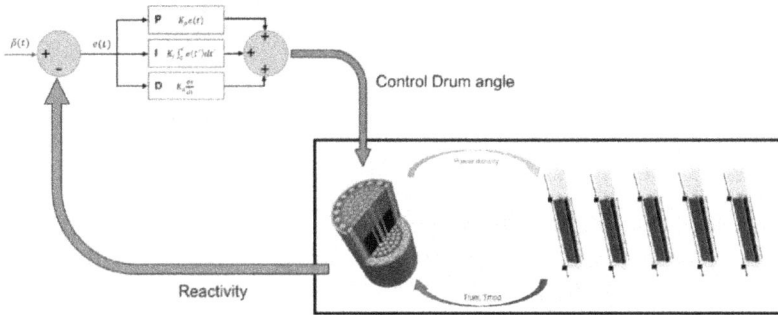

Figure 10.
Schematics of the PID control of the full-core multiphysics model.

corresponds to the reactivity computed by the numerical model with, for instance, an additional typical time delay from the detectors. An error between the desired and measured reactivity is then computed. The updated control drum angle is determined by adding three terms proportional to: (1) the error to attempt instantaneous correction; (2) the integral of the error to account for any persistent underestimating/overestimating of the desired reactivity; and (3) the derivative of the error to anticipate how it is going to evolve in the near-future and avoid over-correction, with the controlling constants called K_p, K_i, and K_d, respectively.

The reason the reactivity is chosen to control the PID—rather than the power—is that a rotation of the drums induces an immediate reactivity change, whereas the corresponding power response is quite delayed (e.g., one may consider reactivity as being roughly the derivative of the power with respect to time). As such, it results in a much more stable control system. However, measured and desired power can be relatively easily converted to reactivity if the neutronic kinetics parameters of the reactor are well known.

The optimal values of K_p, K_i, and K_d can theoretically be determined if the transfer function for the system is known. However, given the complexity of the multiphysics model, it appears impractical to proceed that way. Instead, their values are chosen based on a semi-empirical approach. In particular, K_p represents the angle by which the drums are to be rotated per amount of reactivity. Fortunately, in most of the realistic operational range of the drums, the reactivity inserted per degree (α) is fairly constant and K_p can be approximately set to $1/\alpha$. If the error consistently lags behind the set point or tends to over-correct, the proper approach is to adjust K_i or K_d. In any event, the values of K_p, K_i, and K_d can be adjusted to make the system more or less responsive.

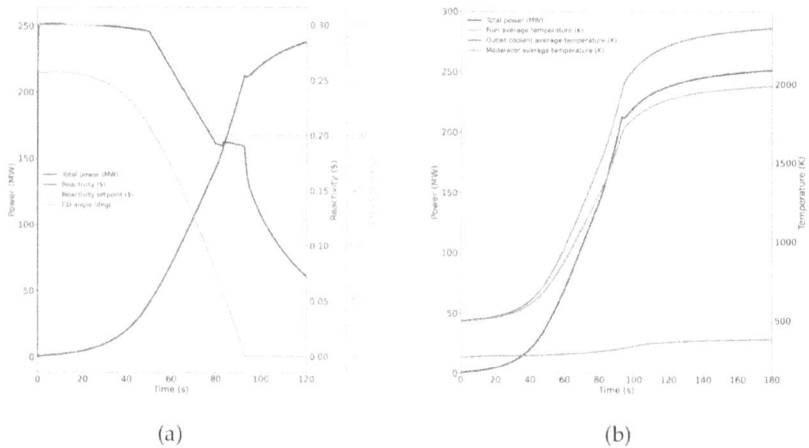

(a)   (b)

Figure 11.
Reactivity setpoint, actual system reactivity, control drum actuation, power response, fuel average temperature, outlet coolant temperature, and moderator average temperature of the generic CERMET NTP system during the startup transient. (a) Reactivity control; (b) Core heating.

In the current simulations, the control system compensates the change in feedback accompanying the change in power well. During the simulation of a startup transient, the reactivity set point is chosen to be 0.3$ for the first 50 s, linearly ramping down to 0.2$ by 80 s of startup, and then remaining constant afterwards. The actual reactivity observed in the simulation closely follows the reactivity set point until the maximum control drum rotation is reached at about 100 s.

Reactor power increases from the initial 610 kW (10 kW per assembly) to close to 250 MW without over-swings in the completed simulation time. At around 90 s, a local maximum in the power is assumed that is attributed to the negative feedback outrunning control drum motion compensating for it. In this case, reactivity is under-compensated.

Temperatures increase monotonically throughout the transient with a corresponding temperature rise in the fuel, and outlet hydrogen being the largest at about 1500 K and moderator temperature rise being very small at less than 120 K. Increase in power will likely have to occur quicker in some NTP operational scenarios. It remains to be investigated if temperatures remain monotonic in these scenarios. The Griffin/RELAP-7/MOOSE model described herein is well equipped to investigate these scenarios (**Figure 11**).

5. Conclusions

This chapter has provided an overview of the concept of nuclear thermal propulsion for interplanetary travel. Nuclear thermal propulsion has a significant advantage in efficiency over current chemical rocket technologies, providing the opportunity to complete a trip to Mars in half the time previously anticipated, reducing exposure time for the spaceship crew. It also offers more options for mission abort if needed.

NTP was first conceived shortly after the end of World War II. Materials development programs and construction and operation of experimental facilities began in the 1950s under Project Rover, which was taken over by NASA shortly after its formation. Rover served as the basis for NASA's Nuclear Engine for Rocket Vehicle

Application (NERVA) program under Werner von Braun, and was planned to enable a mission to Mars, launching in the early 1980s. NERVA ended after funding cuts at the end of the Cold War and the corresponding reduction of the scope of the space program. However, this work was resurrected in the mid 2010s as part of NASA's Game Changing Technology for Deep Space Exploration Program. Much of the experience gained under NERVA was used as a basis for a path forward.

Under NERVA, fuel forms were primarily composed of graphite fuel compacts, although independent work at ANL and GE began developing CERMET fuels. Recent efforts picked up CERMET fuel development, building on the earlier work in addition to other research related to application in other reactor types. NASA also began the evaluation of CERCER fuel forms; all current development efforts are based on the use of HALEU fuel instead of the HEU fuel used within the NERVA program. Tests of fabrication processes and high temperature operation in reactor and non-reactor facilities are underway.

By using the Griffin reactor multiphysics application coupled with the RELAP-7 thermal-fluids systems code and the MOOSE framework, tightly coupled multiphysics simulations are being performed for CERMET-based core designs. The simulation of experiments being performed at the TREAT facility is also underway to aid in the experimental design. Data from the completed experiments are being used to validate the coupled approach.

Much work remains to be completed, both in core design analysis and materials testing to be able to build a prototype nuclear thermal rocket engine. NASA currently plans to launch a manned mission to Mars in 2039. According to a study commissioned by the National Academies of Sciences, Engineering, and Medicine [46], under such an aggressive time schedule, NTP development faces four major challenges: (1) the development of an NTP system that can heat its propellant to approximately 2700 K, which is the core exit for the duration of multiple burn cycles; (2) the need to rapidly bring an NTP system to full operating temperature in a very short time (e.g., on the order of a minute); (3) the long-term storage of LH_2 with minimal loss during a mission; and (4) the lack of U.S. testing facilities for system testing.

Acknowledgements

This work was funded under U.S. Department of Energy contract number DE-AC07-05ID14517, managed by Battelle Energy Alliance, LLC/Idaho National Laboratory for NASA's Space Nuclear Propulsion (SNP) project in the Space Technology Mission Directorate (STMD).

Nomenclature

α	Control drum reactivity inserted per degree of rotation [$1/°$]
c_p	Specific heat specific heat at constant pressure [$J/(kg \cdot K)$]
c_v	Specific heat specific heat at constant volume [$J/(kg \cdot K)$]
g_o	Gravitational constant on earth [m/s^2]
k	Ratio of c_p to c_v for propellant
K_p	PID proportional constant [$°$]
K_i	PID integral constant [$°/s$]
K_d	PID derivative constant [$° \cdot s$]
I	Total impulse [$N \cdot s$]

I_{sp}	Specific Impulse [s]
F_{thrust}	Force (thrust) exerted by propellant [N]
M	Molecular weight of propellant [g/mol]
m_{ex}	Total mass expelled over specific time [kg]
$\dot{m}$	Mass flow rate [kg/s]
T_c	Reactor core exit temperature for NTP or combustion chamber temperature for a chemical engine [K]
p_c	Core exit (or combustion chamber) pressure [N/m^2]
p_e	Nozzle exit pressure [N/m^2]
R	Univeral gas constant [J/kg · mol]
v_e	Exit velocity of propellant relative to nozzle [m/s]
W	Weight on earth [N]

Abbreviations

ATR	Advanced Test Reactor
ANL	Argonne National Laboratory
AEC	Atomic Energy Commission
CERCER	Ceramic–Ceramic
CERMET	Ceramic-Metal
GE	General Electric
GMWDF	Graphite matrix with dispersed fuel
HALEU	High Assay Low Enrichment Uranium
HEU	High Enrichment Uranium
HLLC	Harten, Lax, and Van Leer Contact
INL	Idaho National Laboratory
LASL	Los Alamos Scientific Laboratory
LEO	Low Earth Orbit
LEU	Low Enrichment Uranium
NASA	National Aeronautics and Space Administration
M&S	Modeling and simulation
MTFR	Moderator-to-fuel density ratio
MOOSE	Multiphysics Object Oriented Simulation Environment
NERVA	Nuclear Engine for Rocket Vehicle Application
NRC	Nuclear Regulatory Commission
NTP	Nuclear Thermal Propulsion
NTREES	Nuclear Thermal Rocket Element Environmental Simulator
PDE	Partial differential equation
PID	Proportional-Integral-Derivative
PRIME	Prototypic Reactor Irradiation for Multicomponent Examination
TREAT	Transient Reactor Test facility

Author details

Mark D. DeHart*, Sebastian Schunert and Vincent M. Labouré
Idaho National Laboratory, Idaho Falls, Idaho, United States of America

*Address all correspondence to: mark.dehart@inl.gov

IntechOpen

References

[1] Levack D, Horton J, Joyner C, Kokan T, Widman FW, Guzek B. Mars NTP architecture elements using the Lunar Orbital Platform-Gateway. In: Proceedings of the 2018 AIAA SPACE and Astronautics Forum and Exposition; 17–19 September 2018; Orlando, FL, USA. DOI: 10.2514/6.2018-5106

[2] National Aeronautics and Space Administration. Specific Impulse [Internet]. 2021. Available from: https://www.grc.nasa.gov/www/k-12/airplane/specimp.html [Accessed: 23 June 2021]

[3] Borowski SK. Nuclear thermal propulsion (NTP). In: Blockley R, Shyy W, editors. Encyclopedia of Aerospace Engineering. Available from. DOI: 10.1002/9780470686652.eae115 Accessed: 23 June 2021

[4] Joyner CR et al. LEU NTP engine system trades and mission options. Nuclear Technology; 2020;**206**(8): 1140-115. DOI: 10.1080/00295450.2019.1706982

[5] World Nuclear News. USNC subsidiary supporting cislunar rocket contractors [Internet]. 2021. Available from: https://world-nuclear-news.org/Articles/USNC-subsidiary-supporting-cislunar-rocket-contrac [Accessed: 23 June 2021]

[6] Stewart ME. A historical review of Cermet fuel development and the engine performance implications. In: Nuclear and Emerging Technologies for Space (NETS), 23 February 2015. Albuquerque, NM, USA. 2015. Available from: https://ntrs.nasa.gov/api/citations/20150002852/downloads/20150002852.pdf Accessed: 9 July 2021

[7] Emrich WJ Jr. Principles of Nuclear Rocket Propulsion. Oxford, UK: Butterworth-Heinemann; 2016

[8] Blagonravov AA, editor. Collected Works of K. E. Tsiolkovsky, Vol II: Reactive Flying Machines. A translation of "K. E. Tsiolkovskiy, Sobraniye Sochineniy, Tom II. Reaktivnyye Letatel' nyye Apparaty," Izdatel' stvo Akademii Nauk SSSR, Moscow, 1954. National Aeronautics and Space Administration, NASA TT F-237; Washington, DC. 1965

[9] Bennet J. NASA's nuclear thermal engine is a blast from the Cold War past. Popular Mechanics. 2018. Available from: https://www.popularmechanics.com/space/moon-mars/a18345717/nasa-ntp-nuclear-engines-mars/ Accessed: 21 September 2021

[10] National Aeronautics and Space Administration. Nuclear Thermal Propulsion: Game Changing Technology for Deep Space Exploration [Internet]. 2018. Available from: https://www.nasa.gov/directorates/spacetech/game_chang ing_development/Nuclear_Thermal_Propulsion_Deep_Space_Exploration [Accessed: 21 September 2021]

[11] Ruark AE, editor. Nuclear Powered Flight. APL/JEU-TG-20. Applied Physics Laboratory; Laurel, Virginia. 1947

[12] Spence RW. The Rover nuclear rocket program. Science. 1968;**160** (3831):953-959

[13] Houts MG, et al. NASA's nuclear thermal propulsion project. In: AIAA SPACE 2015 Conference and exposition. Pasadena, CA, USA, 31 August–2 September 2015. Available from: https://arc.aiaa.org/doi/abs/10.2514/6.2015-4523 [Accessed: 4 August 2021]

[14] DeHart MD, Karriem Z, Pope MA. Evaluation of the Enhanced LEU Fuel (ELF) design for conversion of the advanced test reactor to a low-enrichment fuel cycle. Nuclear Technology;**201**(3):247-266. DOI: 10.1080/00295450.2017.1322451

[15] Benensky K. Summary of Historical Solid Core Nuclear Thermal Propulsion Fuels. Report. Pennsylvania: The Pennsylvania State University; 2013. Available from: https://scholarsphere. psu.edu/resources/f44f5ad8-913d-4a0f-a8d4-337b29c6e021/downloads/479 Accessed: 2 September 2021

[16] Finseth JJ. Overview of Rover engine tests. Final report. Report. NASA-CR-184270. 1991. Available from: https://ntrs.nasa.gov/api/citations/ 19920005899/downloads/19920005899. pdf [Accessed: 2 September 2021]

[17] Koenig DR. Experience Gained from the space nuclear rocket program (Rover). Report. LA-10062-H. 1986. Los Alamos, New Mexico: Los Alamos National Laboratory; 1986. Available from: https://nuke.fas.org/space/la-10062.pdf [Accessed: 24 March 2022]

[18] Gaffin ND, Zinkle SJ, Palomares K. Review of irradiation hardening and embrittlement effects in refractory metals relevant to nuclear thermal propulsion applications. In: Nuclear and Emerging Technologies for Space (NETS-2019). American Nuclear Society Topical Meeting. Richland, WA, USA. 25–28 February 2019.

[19] Tucker DS. Cermets for use in nuclear thermal propulsion. In: Lucan D, editor. Advances in Composite Materials Development. London, UK: IntechOpen; 2019. DOI: 10.5772/ intechopen.85220

[20] Burns D, Johnson S. Nuclear thermal propulsion reactor materials. In: Tsvetkov PV, editor. Nuclear Materials. London, UK: IntechOpen; 2020. DOI: 10.5772/ intechopen.91016. Available from: https:// www.intechopen.com/chapters/71396

[21] Schnitzler BG. Small Reactor Designs Suitable for Direct Nuclear Thermal Propulsion: Interim Report. Idaho Falls, Idaho: Idaho National Laboratory: 2012. DOI: 10.2172/1042384

[22] General Electric. 710 High-Temperature Gas Reactor Program Summary Report, Report GEMP-600-V1. Vol. I. Cincinnati, Ohio, USA: General Electric Co., Nuclear Materials and Propulsion Operation; 19681 DOI: 10.2172/4338293

[23] Venneri P, Kim YH, Howe S. Neutronics study on LEU nuclear thermal rocket fuel options. In: Proceedings of the KNS 2014 Fall Meeting. p. 1CD-ROM. Republic of Korea: KNS; 2014

[24] Tonks MR et al. Development of a multiscale thermal conductivity model for fission gas in UO_2. Journal of Nuclear Materials. 2016;**469**:89-98

[25] Ross SB, El-Genk MS, Matthews RB. Thermal conductivity correlation for uranium nitride fuel between 10 and 1923 K. Journal of Nuclear Materials. 1988;**151**(3):318-326

[26] Bhattacharyya, SK. An assessment of fuels for nuclear thermal propulsion. In: ANL/TD/TM01-22. Argonne, IL, USA: Argonne National Laboratory. December 2001. Available from: https:// www.osti.gov/servlets/purl/822135 [Accessed: 4 August 2021]

[27] Argonne National Laboratory. Nuclear Rocket Program Terminal Report. ANL-7236. Lemont, Illinois: Argonne National Laboratory; 1966

[28] Nuclear Energy International. US NRC Approves Licence Amendment for Centrus to Produce HALEU [Internet]. 2021. Available from: https://www. neimagazine.com/news/newsus-nrc-approves-licence-amendment-for-centrus-to-produce-haleu-8827607 [Accessed: 6 August 2021]

[29] Houts M et al. Nuclear cryogenic propulsion stage. In: Nuclear and Emerging Technologies for Space (NETS-2014). Center, Mississippi: NASA Stennis Space; Feb. 24-26, 2014.

Available from: https://ntrs.nasa.gov/api/citations/20140012915/downloads/20140012915.pdf Accessed: 4 August 2021

[30] Schwartzberg FR, Ogden HR, Jaffee RI. Ductile-Brittle Transition in the Refractory Metals. Columbus, Ohio: Battelle Memorial Institute: Defense Metals Information Center; 1959. Available from: https://www.google.com/books/edition/Ductile_brittle_Transition_in_the_Refrac/eElxlohi01wC?hl=en&gbpv=1&pg=PP3 [Accessed: 24 March 2022]

[31] Gaffin N, Ang C, Milner J, Palomares K, Zinkle S. Fabrication of MO30W based cermets for nuclear thermal propulsion using spark plasma sintering. In: Nuclear and Emerging Technologies for Space (NETS 2021). American Nuclear Society Topical Meeting, Las Vegas, NV, USA. 26 February–1 March 2018.

[32] Poston D. Design comparison of nuclear thermal rocket concepts. In: Nuclear and emerging Technologies for Space (NETS 2018). American Nuclear Society Topical Meeting; Oak Ridge, TN, USA; 26–30 April 2021

[33] Youinou GJ, Lin CS, Abou Jaoude A, Casey JJ. Nuclear thermal propulsion scoping analysis of fuel plate core configurations. In: INL/EXT-19-57004. Idaho Falls, ID, USA: Idaho National Laboratory; 2020. DOI: 10.2172/1596107

[34] Gustafson JL. Space nuclear propulsion fuel and moderator development plan conceptual testing reference design. Nuclear Technology. 2021;**207**(6):882-884. DOI: 10.1080/00295450.2021.1890991

[35] Braun R et al. Space nuclear propulsion for human Mars exploration. In: NASEM Space Nuclear Propulsion Technologies Committee Report. Washington, DC: National Academies of Sciences, Engineering and Medicine; 2021. DOI: 10.17226/25977

[36] Benensky K. Tested and Analyzed Fuel Form Candidates for Nuclear Thermal Propulsion Applications. Nuclear Engineering Reports: University of Tennessee; 2016. Available from: https://trace.tennessee.edu/utne_reports/5 Accessed: 24 August 2021

[37] Poston D. Nuclear testing and safety comparison of nuclear thermal rocket concepts. In: Nuclear and emerging Technologies for Space (NETS 2018). Las Vegas, NV, USA: American Nuclear Society Topical Meeting; 26 Feb–1 Mar, 2018

[38] Duderstadt JJ, Hamilton LJ. Nuclear Reactor Analysis. Wiley: New York, NY, USA; 1976

[39] Proust E. Lecture Series on Space Nuclear Power and Propulsion Systems-2-Nuclear Thermal Propulsion Systems (last updated in January 2021). DOI: 10.13140/RG.2.2.17073.10089

[40] Black DL. Consideration of low enriched uranium space reactors. In: AIAA 2018-4673. AIM Propulsion and Energy Forum, July 9-11, 2018, Cincinnati, Ohio: 2018 Joint Propulsion Conference; July 9-11, 2019

[41] Bae IH, Na MG, Lee YJ, Park GC. Calculation of the power peaking factor in a nuclear reactor using support vector regression models. Annals of Nuclear Energy. 2008;**35**(12):2200-2205

[42] Zabriskie AX et al. A coupled multiscale approach to TREAT LEU feedback modeling using a binary-collision Monte-Carlo–informed heat source. Nuclear Science and Engineering. 2019;**193**(4):368-387. DOI: 10.1080/00295639.2018.1528802

[43] Schnitzler B, Borowski S, Fittje J. A 25,000-lbf thrust engine options based

on the small nuclear rocket engine design. In: AIAA 2009-5239. 45th AIAA/ASME/SAE/ASEE Joint Propulsion Conference and Exhibit. Denver, Colorado. August 2009

[44] Fittje JE, Borowoski SK, Schnitzler BG. Revised point of departure design options for nuclear thermal propulsion. In: AIAA 2015-4547. AIAA SPACE 2015 Conference and Exposition, Pasadena, CA, USA. August 2015. Available from: https://arc.aiaa.org/doi/10.2514/6.2015-4547 [Accessed: 10 August 2021]

[45] Nam SH et al. Innovative concept for an ultra-small nuclear thermal rocket utilizing a new moderated reactor. Nuclear Engineering and Technology. 2015;47(6):678-699

[46] National Academies of Sciences, Engineering, and Medicine. Space Nuclear Propulsion for Human Mars Exploration. Washington, DC, USA: The National Academies Press; 2021. DOI: 10.17226/25977

[47] Los Alamos National Laboratory. N-Division Personnel. Pewee I Reactor Test Report. Los Alamos National Laboratory Informal Report, LA-4217. Los Alamos, NM, USA: Los Alamos National Laboratory; 1969

[48] Stacy WM. Nuclear Reactor Physics. New York, NY, USA: Wiley; 2007

[49] Gates JT, Denig A, Ahmed R, Mehta VK, Kotlyar D. Low-enriched cermet-based fuel options for a nuclear thermal propulsion engine. Nuclear Engineering and Design. 2018;331:313-330. DOI: 10.1016/j.nucengdes.2020.110605

[50] Krecicki M, Kotlyar D. Low enriched nuclear thermal propulsion neutronic, thermal hydraulic, and system design space analysis. Nuclear Engineering and Design. 2020;363:110605

[51] Walton JT. An overview of tested and analyzed NTP concepts. NASA Technical Memorandum 105252, AIAA-91-3503, Conference on Advanced Space Exploration Initiative Technologies, cosponsored by AIAA, NASA, and OAI, Cleveland, Ohio, September 4-6, 1991. Available from: https://ntrs.nasa.gov/api/citations/19920001919/downloads/19920001919.pdf [Accessed: 24 March 2022]

[52] Permann CJ et al. MOOSE: Enabling massively parallel multiphysics simulation. SoftwareX. 2020;11:100430. Available from: https://www.scienced irect.com/science/article/pii/S2352711019302973 [Accessed: 24 March 2022]

[53] Kirk BS, Peterson JW, Stogner RH, Carey GH. libMesh: A C++ library for parallel adaptive mesh refinement/coarsening simulations. Engineering with Computers. 2006;22(3–4):237-254. DOI: 10.1007/s00366-006-0049-3

[54] Balay S, et al. PETSc Users Manual. ANL-95/11 Rev 3.7. 2016. Available from: https://ntrs.nasa.gov/api/cita tions/20140012915/downloads/20140012915.pdf [Accessed: 1 November 2021]

[55] Martineau R et al. Multiphysics for nuclear energy applications using a cohesive computational framework. Nuclear Engineering and Design. 2020;367:110751. DOI: 10.1017

[56] Wang Y et al. Rattlesnake: A MOOSE-based multiphysics multischeme radiation transport application. Nuclear Technology;207(7):1047-1072. DOI: 10.1080/00295450.2020.1843348

[57] Wang Y et al. Performance improvements for the Griffin transport solvers. In: INL/EXT-21-64272-Rev000. Idaho Falls, ID, USA: Idaho National Laboratory; 2021. Available from: https://inldigitallibrary.inl.gov/sites/sti/sti/Sort_50897.pdf [Accessed: 15 September 2021]

[58] Shemon ER et al. PROTEUS-SN User Manual, Revision 1.2. ANL/NE-14/6. Chicago, IL, USA: Argonne National Laboratory; 2014

[59] Lee CH et al. MC2-3: Multigroup cross-section generation code for fast reactor analysis. In: ANL/NE-11/41, Rev. 3. Chicago, IL, USA: Argonne National Laboratory; 2018

[60] Schunert S et al. Control rod treatment for FEM based radiation transport methods. Annals of Nuclear Energy. 2019;**127**:293-302

[61] Williamson RL et al. BISON: A flexible code for advanced simulation of the performance of multiple nuclear fuel forms. Nuclear Technology;**207**(7): 954-980. DOI: 10.1080/00295450. 2020.1836940

[62] Hirschhorn J et al. Review and preliminary investigation into fuel loss from cermets composed of uranium nitride and a molybdenum-tungsten alloy for nuclear thermal propulsion using mesoscale simulations. Journal of Materials. 2021;**73**(11):3528-3543. DOI: 10.1007/s11837-021-04873-x. Available from: https://link.springer.com/content/pdf/10.1007/s11837-021-04873-x.pdf [Accessed: 24 March 2022]

[63] Berry RA et al. RELAP-7 Theory Manual. INL/EXT-14-31366, Rev. 2. Idaho Falls, ID, USA. Available from: https://inldigitallibrary.inl.gov/sites/sti/sti/6899506.pdf: Idaho National Laboratory; 2016 Accessed: 15 September 2021

[64] Adhikary D, Jayasundara C, Podgorney R, Wilkins A. A robust return-map algorithm for general multisurface plasticity. International Journal for Numerical Methods in Engineering. 2017;**109**(2):218-234. DOI: 10.1002/nme.5284

[65] Wang Y, et al. Demonstration of MAMMOTH strongly-coupled multiphysics simulation with the Godiva benchmark problem. In: M&C 2017 - International Conference on Mathematics & Computational Methods Applied to Nuclear Science & Engineering, Jeju, Korea, April 16-20, 2017. Available from: https://www.kns.org/files/int_paper/paper/MC2017_2017_9/P353S09-01WangY.pdf [Accessed: 24 March 2022]

[66] Klein AC, Camp AL, PR MC, Voss SS. Operational Considerations for Fission Reactors Utilized on Nuclear Thermal Propulsion Missions to Mars—A Report to the Nuclear Power & Propulsion Technical Discipline Team. NASA Technical Report NASA/CR20210000387. Hampton, VA, USA: Langley Research Center; Jan 2021

[67] Xia Y et al. Preliminary Study on the Suitability of a Second-order Reconstructed Discontinuous Galerkin Method for RELAP-7 Thermal-Hydraulic Modeling. INL/EXT-17-43108-Rev001. Idaho Falls, ID, USA: Idaho National Laboratory; 2017. DOI: 10.2172/1468483

[68] Labouré V et al. Hybrid super homogenization and discontinuity factor method for continuous finite element diffusion. Annals of Nuclear Energy. 2019;**128**:443-454. DOI: 10.1016/j.anucene.2019.01.003

[69] Leppänen J et al. The serpent Monte Carlo code: Status, development, and applications in 2013. Annals of Nuclear Energy. 2015;**82**:142-150. DOI: 10.1016/j.anucene.2014.08.024

Section 2

Research Reactors

Chapter 2

The Transient Reactor Test Facility (TREAT)

Nicolas Woolstenhulme

Abstract

Constructed in the late 1950s, the Transient Reactor Test facility (TREAT) provided numerous transient irradiations until operation was suspended in 1994. It was later refurbished, and resumed operations in 2017 to meet the data needs of a new era of nuclear fuel safety research. TREAT uses uranium oxide dispersed in graphite blocks to yield a core that affords strong negative temperature feedback. Automatically controlled, fast-acting transient control rods enable TREAT to safely perform extreme power maneuvers—ranging from prompt bursts to longer power ramps—to broadly support research on postulated accidents for many reactor types. TREAT's experiment devices work in concert with the reactor to contain specimens, support in situ diagnostics, and provide desired test environments, thus yielding a uniquely versatile facility. This chapter summarizes TREAT's design, history, current efforts, and future endeavors in the field of nuclear-heated fuel safety research.

Keywords: transient testing, fuel safety research, accident simulation

1. Introduction

In the late 1950s, the Transient Reactor Test facility (TREAT) was designed, constructed, and commissioned within the span of only a few years [1]. The facility was built just over 1 km away from the Experimental Breeder Reactor-II (EBR-II) sodium-cooled fast breeder reactor as part of the Argonne National Laboratory West campus (ANL-W) located in the Arco Desert, west of Idaho Falls, Idaho. As with most facilities at ANL-W, TREAT was originally envisioned to help support research and development pertaining to EBR-II, but its mission diversified in later years to support other nuclear technology areas. TREAT was a specialized graphite-based test reactor able to safely perform extreme transient power maneuvers to research the effects of postulated accident conditions on nuclear fuel specimens placed in its core [2, 3]. A modern aerial image of TREAT is shown in **Figure 1**.

TREAT's unique abilities stem from its fuel assemblies, in which uranium oxide, graphite, and carbon powders are mixed with binders, pressed into blocks, and fired at high temperatures [4]. The resulting fuel blocks were stacked inside zircaloy-3 sheet metal canisters (a uniquely oxidation-resistant zirconium alloy that was being researched at the time, but which is no longer in production, having been superseded by other zirconium alloys for light-water reactor [LWR] use). These canisters were evacuated and sealed. Aluminum sheaths and end fitting hardware were fastened to the tops and bottoms of these fuel assemblies to house graphite reflectors and provide mechanical interfaces for gridplate placement and handling. These fuel assemblies had a ~10 cm^2 cross section with 0.6 m of unfueled axial

Figure 1.
Modern day aerial image of TREAT.

reflector top and bottom with 1.2 m of active fueled length in the center. Various special fuel and graphite dummy assemblies were also produced, including some with central cylindrical cavities for control rods, some with integral thermocouples, and some with a void region (i.e., containing no fuel or moderator) in the core's axial center (see **Figure 2**) [5].

The resulting fuel assemblies were produced in sufficient quantity to fill the reactor's 19 × 19 square-pitch gridplate array. Despite thousands of reactor startup and transient cycles over the decades that followed, the fluence experienced during short transients was small, and these same fuel assemblies accumulated very little burnup. Hence, TREAT operates to this day using the original fuel assemblies produced in the 1950s. Occasionally, these fuel assemblies are shuffled into different reactor positions or stored below grade in adjacent storage holes. Core reconfigurations are performed to optimize the core parameters for experimental needs rather than to equilibrate burnup as is typical of most nuclear reactor shuffling schemes. The radionuclide inventory of these fuel assemblies is minimal, and they can be handled without shielding, especially after an extended decay period

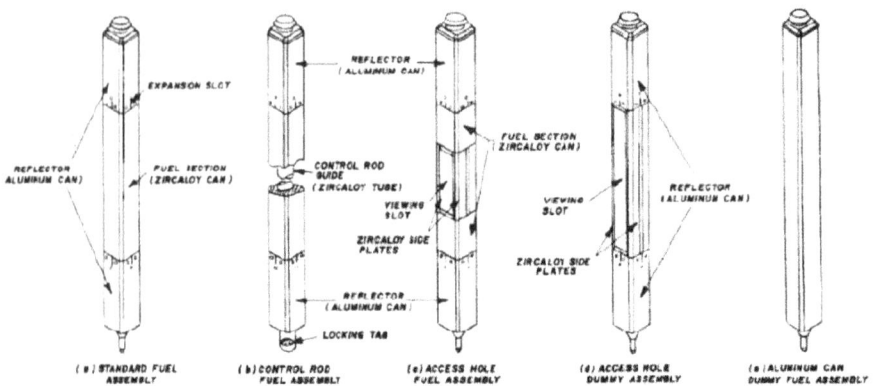

Figure 2.
Historic image of TREAT fuel assembly types.

between transient operations. Still, these fuel assemblies are typically handled in a lead-shielded cask outside the reactor to reduce personnel radiation exposure.

TREAT's active core region resided just above ground level. The reactor is surrounded by a thick wall of graphite reflector blocks. TREAT's graphite reflector is surrounded by thick walls of concrete that comprise both the reactor's main structural shell and its radiation shielding. Blocks can be removed from some parts of the graphite reflector and concrete shielding to create a void slot for viewing the core center from each of the four cardinal directions. Presently, the west slot is occupied by a collimated-beam neutron radiography facility adjacent to the reactor. The north slot is occupied by the Fuel Motion Monitoring System (FMMS), also known as the hodoscope. The east slot area is filled with normal fuel assemblies with a large graphite region in the concrete wall and rolling shield door give access to a highly thermalized neutron environment referred to as the thermal column. The south slot is currently unused, but could be outfitted with other scientific instruments or facilities in the future.

The concrete walls support a ~30 cm-thick circular upper shield plug approximately 3 m in diameter. This shield plug can rotate 360 degrees on bearings via a gear drive. A rectangular slot through the shield plug extends from its center to its periphery. All fuel assemblies, experiments, and other hardware are installed in TREAT through this slot, using bottom loading shielded casks and/or overhead cranes. A ~1 m gap between the top of the fuel assemblies and the bottom of the rotating shield plug provides space for TREAT's control rods to protrude above the core. See **Figures 3** and **4** for an overview of some of the reactor's key features.

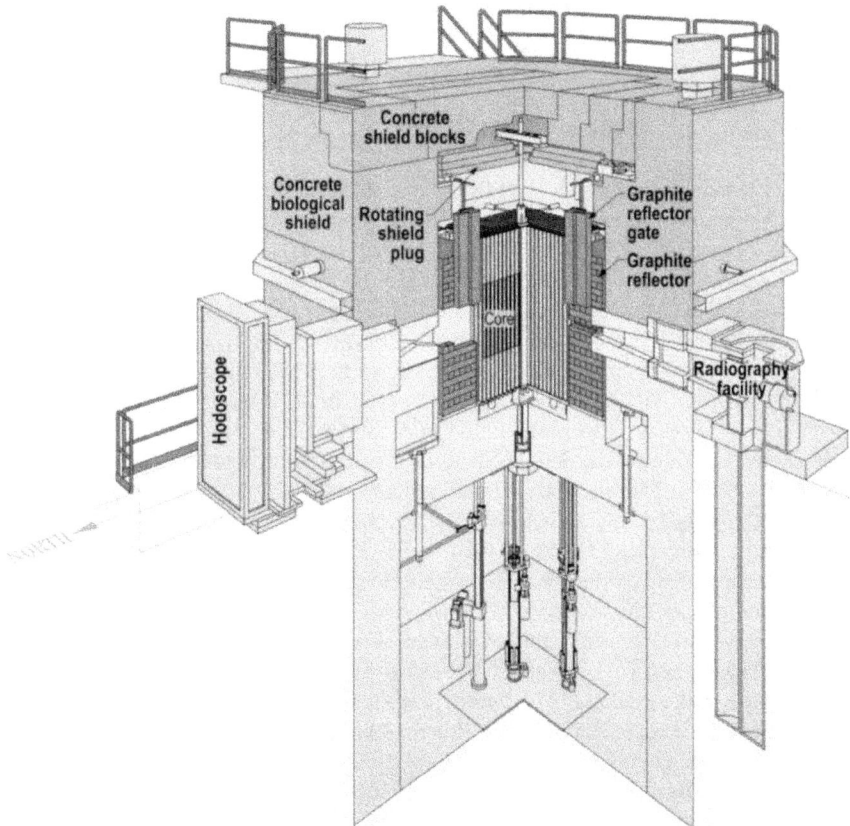

Figure 3.
Section view of TREAT's key reactor features [6].

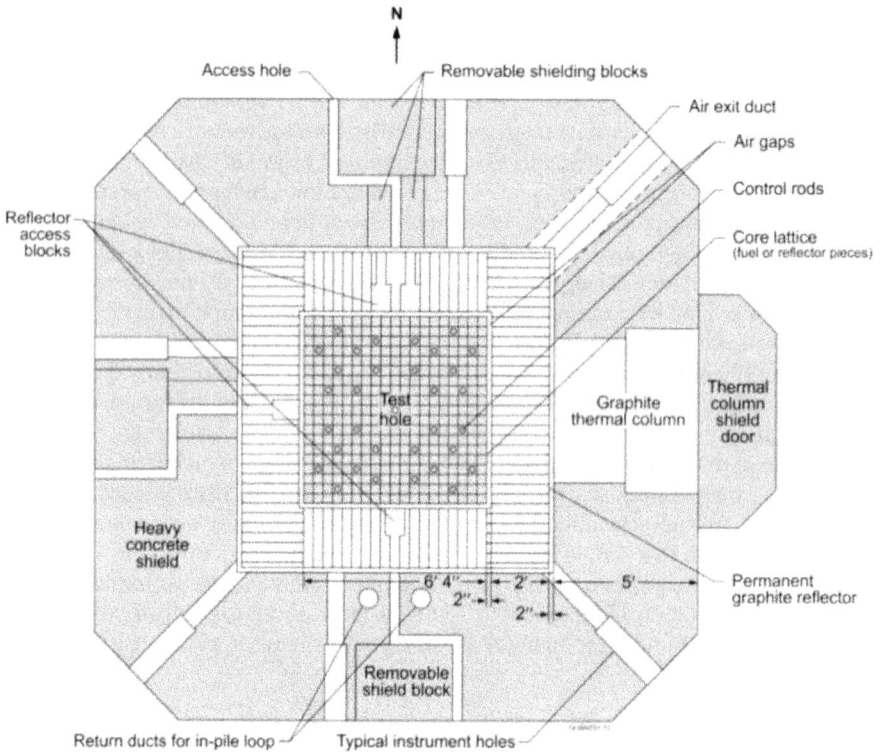

Figure 4.
Top view of TREAT's core, reflector, and shielding layout [6].

Apart from experimental devices that may contain various liquids to support the desired specimen boundary conditions, TREAT does not house liquid coolant for the reactor itself. Instead, a blower system pulls air from the reactor building through debris filters located atop the reactor, down into the core (primarily through ~1 cm^2 coolant channel gaps where the corner chamfers of four fuel assembly canisters meet), and out through a filtration system and stack. This air-cooling system is adequate to enable the reactor to operate in low-level steady-state (LLSS) mode for several hours at a time. Presently, TREAT is authorized to operate in LLSS mode at up to 120 kW thermal power, but this power level does not challenge facility physical limitations and could likely be uprated if needed. LLSS mode is useful for calibrations, system check outs, dosimeter irradiations, and neutron radiography. This cooling system is inadequate for removing significant heat within the time duration of a fast transient; hence, it is not credited for transient safety calculations. Therefore, the core's heat capacity and high-temperature oxidation of fuel assembly canisters typically set the core transient energy capacity at around 2500 MJ, depending on the core configuration. The cooling system also helps cool down the core after large transients, thus boosting operational efficiency. In this manner, TREAT can typically perform one large transient per day—and occasionally two moderate-energy transients in a one-day shift.

TREAT's unique core design is complimented by its specialized control rod systems, thus enabling its unparalleled transient capabilities. All TREAT's control rod types use boron carbide in the absorber section, along with graphite-filled zirconium alloy followers. Reactor operation is initiated by withdrawing compensation and transient rod sets (the compensation rods' purpose is to ensure that hold-down reactivity margins are maintained during the removal of large experiment devices, many of which are net neutron sinks). The reactor is then brought critical by moving the control/shutdown rod sets out of the active core. LLSS operations are typically performed with the rods in this configuration. Transient control rods can then be inserted incrementally to prepare for transient operations, while the control/shutdown rods are withdrawn to maintain criticality until the desired excess reactivity is available in the transient rods. The reactor is then switched into transient mode, and a preprogrammed transient power shape is executed by an automatically controlled computer system with active feedback from ion chamber neutron detectors located in TREAT's concrete shielding. See **Figure 5** for an example core map showing these control rod locations.

Transient rods are driven by fast-acting hydraulics in the TREAT basement sub-pile room (see **Figure 6**). These rod drives can move the rods at a velocity of ~3.5 m/s in both directions (i.e., up and down), permitting split-second manipulation of the reactor's power shape. A tremendous number of transient shapes can be executed, including prompt pulses, ramps, flattop regions, and combinations thereof [7]. (See **Figure 7** for examples of possible power shapes in TREAT.) Transient operation can be "clipped," based on the desired test conditions, by rapidly inserting the transient control rods to narrow the TREAT natural pulse width to <90 ms (full width at half max) and terminate the reactor power. Further upgrades are planned for expanding TREAT's clipping capability to include even narrower pulses when needed [8].

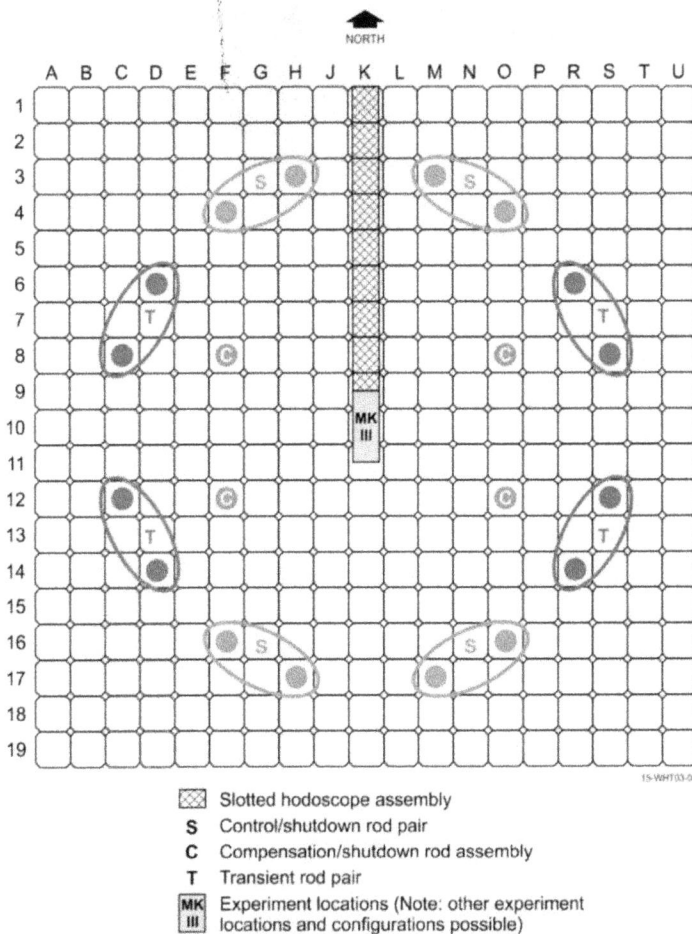

Slotted hodoscope assembly
S Control/shutdown rod pair
C Compensation/shutdown rod assembly
T Transient rod pair
MK III Experiment locations (Note: other experiment locations and configurations possible)

Figure 5.
Example core map showing current control rod types and locations.

Figure 6.
View of TREAT rod drive mechanisms from the sub-pile room.

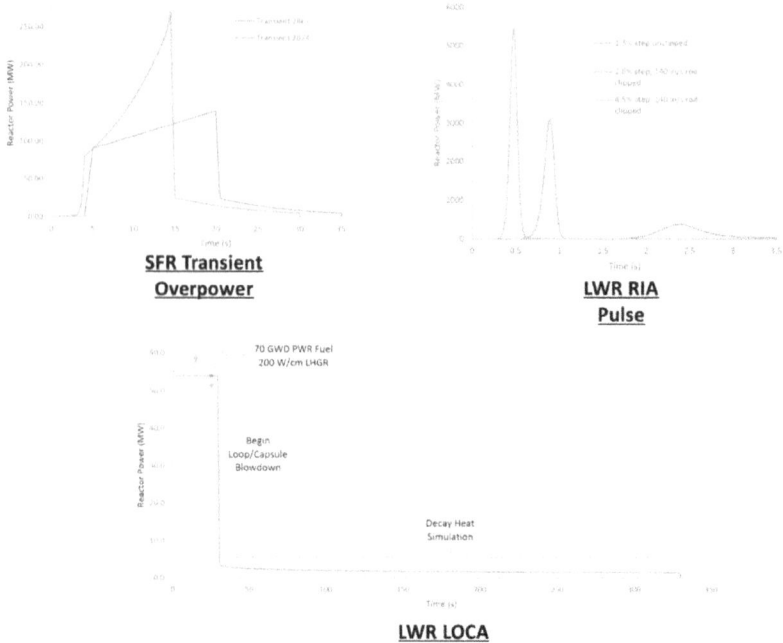

Figure 7.
Example transient shapes possible at TREAT.

Under certain conditions, a state-of-the-art reactor trip system will initiate the rapid insertion of all rods; however, as with the air-cooling system, the trip system is not credited in the reactor safety basis. Instead, TREAT's strong negative temperature feedback behavior is credited as the primary means of limiting transient energy. Since TREAT's uranium oxide particles are dispersed in the fuel blocks, power excursions cause the moderator temperature to rise, resulting in higher neutron energy, increased neutron leakage, and self-limiting power excursions with reliable negative temperature reactivity coefficients. This key feature of TREAT enables it to safely perform research on nuclear fuel specimens under extreme conditions.

2. Facility history

The Arco Desert, where TREAT and ANL-W were built, has also housed many other test reactors as part of the National Reactor Testing station and Naval Reactor Facility missions. A series of water-based transient test reactors were constructed under the Special Power Excursion Reactor Test (SPERT) program that was contemporary to TREAT in its early years [9]. Together, SPERT and TREAT used water capsules to conduct most of the foundational research on overpower fuel performance thresholds for LWRs. During this time, TREAT also continued to perform research on sodium fast reactor (SFR) fuels and nuclear thermal propulsion (NTP) fuels using specialized test capsules. Later, two additional landmark facilities were built out in the Arco Desert to advance research on the accident behavior of LWR fuels. The Power Burst Facility (PBF) offered unrivaled capabilities for reactivity-initiated-accident testing of fuel rods in an integral pressurized flowing loop [10], while the Loss-of-Fluid Test Facility (LOFT) addressed system-scale safety testing via its seminal work in loss-of-coolant-accident testing [11]. These features, along with the postmortem exams performed by facilities in the Arco Desert on fuel from the Three Mile Island accident made Idaho the nexus of fuel safety research throughout the 1980s.

With PBF and LOFT focusing on LWR safety research TREAT's latter historic era naturally shifted toward a focus on SFR fuels using clever sodium loop test vehicles. The Mk-series loops could test bundles of up to seven pins using compact electromagnetic pumps to recirculate sodium through a small pipe weldment [12]. The entirety of these loops was small and self-contained to foster transportation between TREAT and the adjacent Hot Fuel Examination Facility (HFEF) on the main ANL-W campus. Casks established for this purpose could house sodium loops and other experiments measuring up to 25 cm in diameter by 3.6 m tall. HFEF was used to assemble fuel pins irradiated in other test reactors (e.g., EBR-II) into these TREAT test loops, and to extract/examine these pins after transient irradiation. Today, HFEF remains in operation as a global hub for post-irradiation examinations.

Unlike reactors such as PBF, TREAT was not designed from the ground up with integral piping for test loops. Thus, the most common type of TREAT experiment design is well represented by the successful Mk-series sodium loops. Referred to as package- or cartridge-type experiments, this design approach used a compact, robust, experiment containment vessel to provide the desired specimen boundary conditions and contain all chemical, radiological, and mechanical hazards associated with the test (see **Figure 8**). These devices, which fit entirely within casks, were installed by being lowered into the reactor and then connected to power/signal lead on the top flange. These leads were routed through the slot in the rotating shield plug and to the necessary control and data acquisition equipment. The absence of liquid coolant or pressure vessel surrounding the reactor simplified lead routing for facilitating transient tests in which real-time experiment data was crucial for understanding the data objectives. This package-type approach was key for enabling TREAT to address specimen coolant conditions and research needs for a variety of reactor designs [13].

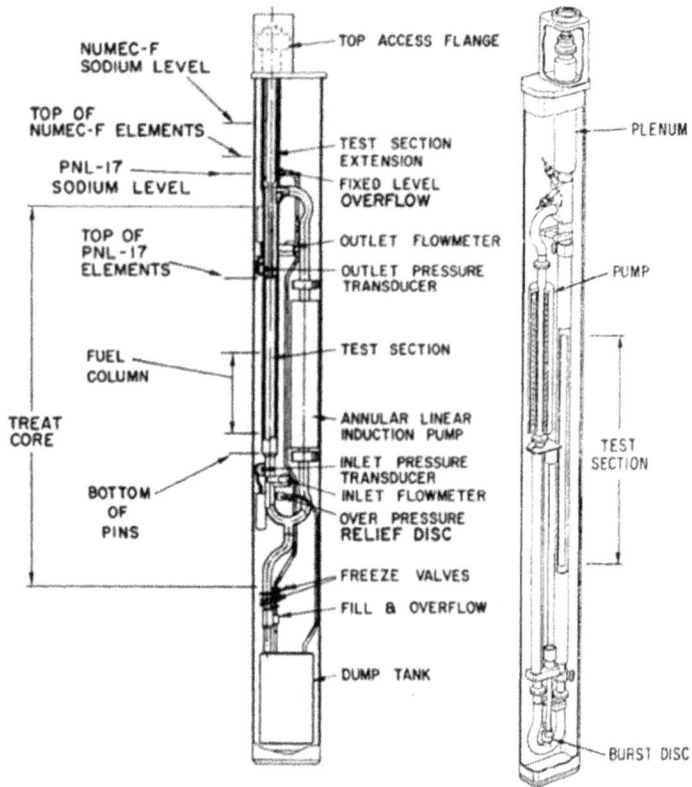

Figure 8.
Historic images of Mk-series sodium loop designs.

TREAT performed numerous tests on oxide-type SFR fuel designs in Mk-series loops to produce much of foundational transient behavior data for these systems. The TREAT facility was upgraded in numerous ways to enable testing of larger oxide fuel bundles in an upsized sodium loop in order to address further data gaps, but shifts in national research priorities prevented this upgrade project from being fully completed. The major upgrades that were realized included a larger building with increased crane capabilities and experiment storage holes, modernization of the automatic reactor control system, and reconfiguration/upgrading of the control rod configuration for the reactor trip system (described earlier). While the upsized sodium loop was never deployed, a special set of new TREAT driver fuel assemblies was also fabricated—using higher uranium loading and Inconel canisters to support higher temperature operation—in an inner converter ring meant to increase the fast neutron flux delivered to the test. These new upgrade driver fuel assemblies remain unused in storage at TREAT to this day.

TREAT was upgraded and maintained in state-of-the-art condition up through the early 1990s. By this time, SPERT, PBF, and LOFT had all ceased operation. TREAT continued to perform work related to SFR metallic fuel until funding was canceled for the Integral Fast Reactor Program in the mid-1990s, causing both TREAT and EBR-II to cease operation. EBR-II was eventually decommissioned, and unique specimens irradiated therein were placed in storage to await future use. However, TREAT's unique, simple design required virtually no maintenance to remain in a safe condition. As a result, electrical power to TREAT's control rod drive systems was simply disconnected to ensure it could not operate, fuel was left in the reactor, and it remained unchanged in this state for approximately 20 years.

Owing to its floorspace, vertical headroom, and authorization as a nuclear facility, TREAT was still used throughout these years for various other nuclear research applications, but the reactor itself was not operated. These efforts required TREAT to remain in active status and maintain its safety basis authorization. Throughout this period, occasional efforts surfaced to champion the resumption of reactor operations at TREAT [14], but none garnered enough momentum to realize this goal. The events of Fukushima Daiichi in 2011, however, gave rise to renewed interest in developing and researching enhanced safety characteristics for nuclear fuels. The U.S. Accident Tolerant Fuels (ATF) program was launched shortly thereafter and, along with the other mission needs that had accumulated over the years, finally justified the resources needed to resume reactor operations at TREAT [15].

The TREAT restart project then followed. The entirety of the TREAT restart project is summarized in a journal special issue in Ref. [16]. Articles from this special issue are referenced throughout this paper as appropriate. The facility was thoroughly characterized and refurbished as needed, with a focus on age-related degradation of systems and components. In some cases, basic industrial equipment in the plant was replaced or repaired, but most of the plant's systems were found in good working order. Key staff previously involved in TREAT operation, many of whom had since retired, rallied to this project to train new staff and transfer knowledge. The facility's safety basis authorization was updated and modernized to reflect new standards and needs [17]. As a testament to the facility's simplicity, the orderly way it was shut down, and the dedication of the restart project staff, TREAT achieved its "second first-criticality" in 2017 [18]—both ahead of schedule and under budget [19].

A few years prior to TREAT's restart, contractor reorganization caused the ANL-W campus, along with its key facilities (i.e., TREAT and HFEF), to come under the same management structure responsible for operating many other key nuclear research assets, including the Advanced Test Reactor (ATR). The resulting national laboratory was termed the Idaho National Laboratory (INL). Upon TREAT's successful restart, INL attained a powerful partnership in research reactor facilities (e.g., a high-flux thermal spectrum material test reactor [ATR], a multipurpose transient test reactor [TREAT], and a sizeable hot cell with abilities to examine and transfer specimens between these reactors [HFEF]) (**Figure 9**).

Figure 9.
Modern-day aerial view of INL's materials and fuels complex. (ATR is just out of view on the left side, ~30 km west of TREAT.)

3. Current efforts and future outlook

Efforts to prepare for transient experiments began shortly after the TREAT restart project commenced. The FMMS detectors were refurbished, and its data acquisition system was replaced with a modern digital system at this time. The FMMS works as fast neutrons born in the experimental fuel specimens travel through the experiment's containment structure, the core's void "slotted" assemblies, and one of several slits in a collimator installed in the reactor's concrete shielding. A fast neutron detector resides at the end of each slit. The slits are arrayed to focus on different axial and transverse locations in the experiment cavity. The FMMS detectors interact with fast neutrons to cause scintillation and luminescence. This phenomenon is proportional to the number of fast neutron interactions, becomes amplified by photomultiplier tubes, and is converted into an electrical signal for high-speed digital data acquisition. This FMMS is able to observe the location of test fuel throughout the duration of the transient. Phenomena such as the expansion, disruption, and meltdown of test fuel can be observed in real time by the FMMS. A cross-section image of the FMMS can be seen in **Figure 10**.

Similarly, the new digital experiment data acquisition and control system (EDACS) was installed. EDACS relies on commercially available equipment and is designed with modularity and expandability to support new instrumentation and control system functions. Dedicated controllers work redundantly with this system to ensure that functions significant to safety are highly reliable (e.g., overtemperature control of electric heaters for heating experiments prior to transient operation). Similarly, wire routing options and facility locations were established for special-purpose signal processing and data acquisition equipment to support special test sensors that do not require integration with EDACS.

In the years preceding its restart, numerous experimenters had expressed interest in using TREAT. The interests of these users encompassed LWR-, SFR-, and NTP-type reactors. A new test system, referred to as the Minimal Activation Retrievable Capsule Holder (MARCH), was designed to fulfill these various research needs shortly after resuming reactor operations. The MARCH system took inspiration from historic package-type experiments by using a stainless-steel containment pipe weldment, inside a sheet metal enclosure, referred to as the Broad

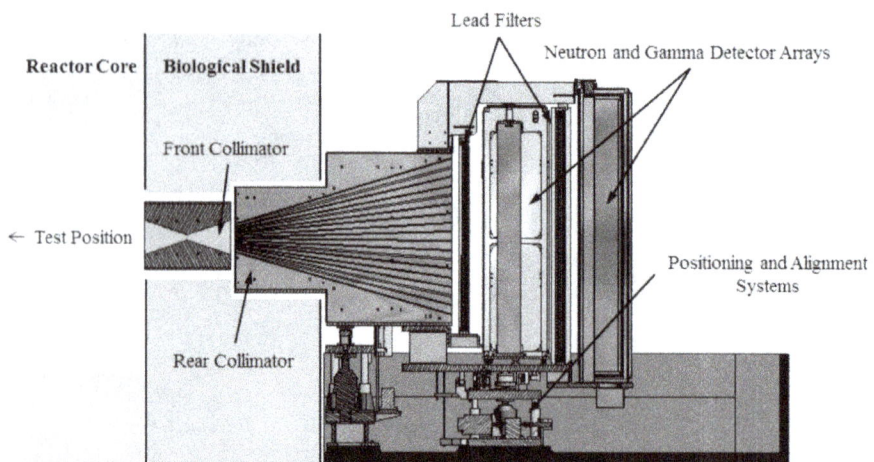

Figure 10.
FMMS cross section of collimator/detector locations and the reactor shielding interface [20].

Use Specimen Transient Experiment Rig (BUSTER). BUSTER can be handled, installed, and connected to support leads in the same way as the Mk-series loops. However, the MARCH system departed from the historic approach in that the sealed capsules are placed inside its pipe. Since many of the first fuel technologies tested in TREAT were emerging (e.g., ATF specimens), only fresh fuel specimens were available. Hence, by combining this capsule-in-pipe mechanical layout with capsule materials that do not transmute into significant radioisotopes (principally titanium alloys), the MARCH system enabled fresh fuel capsules to be irradiated, removed from BUSTER on the TREAT working floor by using the storage holes, and shipped for post-transient exams using glovebox facilities, all in a matter of weeks. A detailed characterization of the BUSTER nuclear environment was performed via Monte Carlo neutronics modeling and can be found in [21]. This approach enables BUSTER to function as a reusable device manufactured in accordance with exacting pressure vessel code and quality assurance requirements, whereas capsules are typically treated as consumable hardware with function-specific engineering requirements. This strategy helps reduce costs as well as the design innovation cycles between test series and capsule adaptations.

The inaugural irradiations performed in BUSTER were sponsored by the ATF program and featured LWR rodlets composed of UO_2 pellets in zirconium-alloy cladding. These tests used a helium environment capsule design known as the Separate Effects Test Holder (SETH). These tests focused on quantifying core-to-specimen energy coupling factors, commissioning new experiment support systems such as EDACS, demonstrating use of the FMMS, and assessing the performance of instrumentation in concurrent tests placed in TREAT coolant channel positions [22]. The SETH tests hosted new technologies for world first applications in transient testing, including additively manufactured capsules and multispectral pyrometry. Post-transient exams were performed as intended using a glovebox facility [23], and a second round of capsules were irradiated shortly thereafter on ATF technologies including as U_3Si_2 fuel pellets and silicon carbide composite cladding [24]. The design was adapted to perform power ramp testing on unclad ceramic fuel specimens inside solid metal holders acting as heat sinks to create thermomechanical gradients in order to investigate transient fuel fracture behaviors.

Building on the successes of the SETH series of experiments, three new major capsule categories were created to provide more prototypic specimen boundary conditions. One capsule was created to support new NTP fuel specimen testing in the SIRIUS series of experiments. The SIRIUS capsule design can house hydrogen in its gas environment, as well as support repeated high-temperature irradiations. The SIRIUS capsule has been used to perform repeated power ramps and to measure specimen temperatures ranging from room temperature to well beyond 2000°C in order to simulate NTP engine startup cycles.

Another capsule, termed the Static Environment Rodlet Transient Test Apparatus (SERTTA), was created to house pressurized water environments for reactivity-initiated-accident testing on LWR rodlets. To date, several studies have been performed using SERTTA, including a series of tests focused on the elucidation of in-reactor transient critical heat flux boiling behavior, and aided by a novel electro-impedance sensor able to detect water voiding in real time [25]. The SERTTA capsule was also recently used to test an LWR rodlet previously irradiated in the ATR. This test marked the first modern use of HFEF to assemble TREAT experiments. Tests assembled in HFEF are expected to become prevalent as more previously irradiated specimens become available for end-of-life fuel safety testing.

A new sodium capsule, referred to as the Temperature Heat-Sink Overpower Response (THOR) capsule, was very recently designed and underwent commissioning tests in TREAT. THOR's key feature is a thick-walled metal heat sink surrounding the specimen. Embedded electrical heaters liquify sodium between the heat sink and test pin cladding prior to transient operation. The liquid sodium enables tight thermal coupling between the pin and heatsink. Working in concert with TREAT's flexible transient power-shaping capability, THOR can simulate transient overpower temperature responses in test pins. THOR can house up to a single full-length EBR-II rod and is currently being prepared for a test series using legacy rods irradiated in EBR-II that were retained for many decades for this very purpose. See **Figure 11** for an overview of the test capsules currently used in the MARCH system at TREAT.

As of 2021, TREAT offers a variety of experiment capabilities and capsules for testing fuel specimens in water, liquid metal, inert gas, and NTP reactor environments. As the only remaining U.S. transient test reactor with significant fuel testing capabilities, TREAT's mission in the modern era remains as diverse as ever. Still, TREAT and its supporting infrastructure are not yet as capable as they were in the past, especially considering that TREAT must now absorb missions that would historically have been addressed by other reactors. This need is particularly important for test devices able to house larger specimens/bundles and actively manipulate thermal hydraulic conditions. For this reason, a new enlarged version of BUSTER (i.e., Big-BUSTER) has been engineered and slated for deployment in TREAT in 2022. Big-BUSTER allows for test devices up to 20 cm in diameter (as opposed to the 6 cm

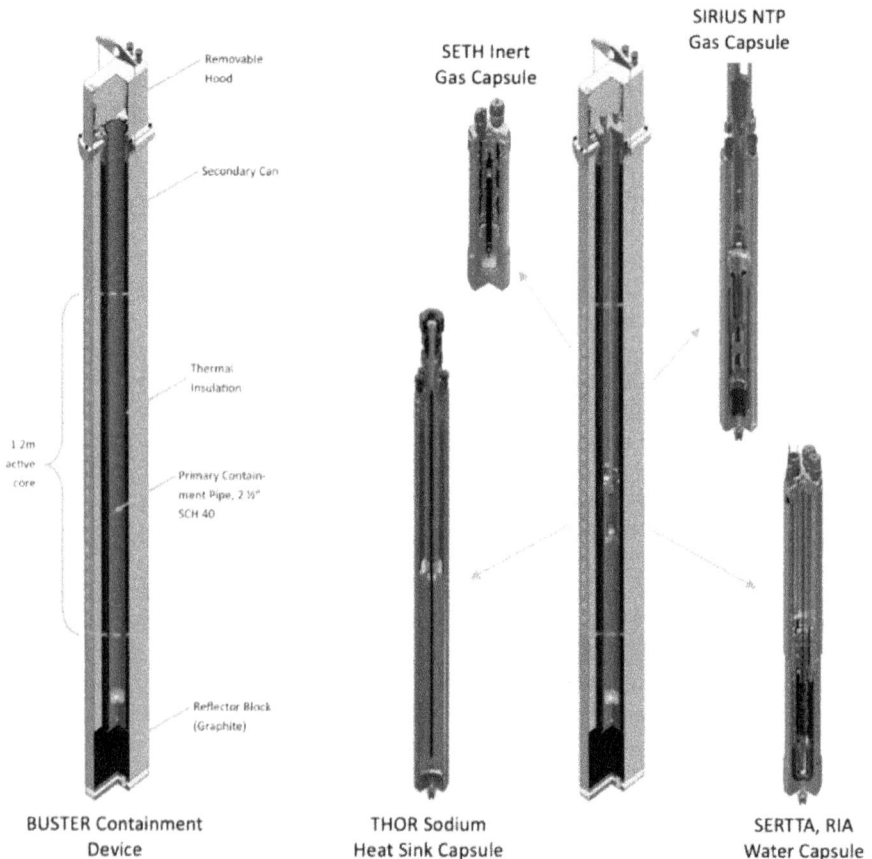

Figure 11.
Overview of MARCH system and experiment capsules used to date.

available in BUSTER), and is constructed from a zirconium alloy to afford increased neutron flux to the test device.

Currently, Big-BUSTER is planned to house an enhanced pressurized water capsule. This capsule is based on a design originally intended to fit in BUSTER with a water blowdown tank to simulate LWR loss-of-coolant accidents [26] but enlarged and adapted to Big-BUSTER for larger test rods. Hot-cell-based equipment is currently under development to enable full-length LWR rods to be cropped, rewelded/pressurized, and outfitted with instrumentation to support such tests. The historic Mk-series sodium loop was also updated to feature modern components and adapted to fit within Big-BUSTER. This new sodium loop will be used to irradiate SFR specimens and small bundles, including longer pins historically irradiated in the now-decommissioned Fast Flux Test Facility. These pins were shipped to INL decades ago and retained for many years to address transient data needs. Other test devices currently under development involve plans to use Big-BUSTER for enhanced test environment simulation. Notable projects planned for deployment include a flowing hydrogen loop for testing advanced NTP fuels, and a helium gas-cooled device for testing microreactor and other gas-cooled reactor technologies. Based on this trajectory, TREAT is expected to continue expanding its capabilities and missions to likely become the longest lived and most versatile transient test reactor ever constructed.

Author details

Nicolas Woolstenhulme
Idaho National Laboratory, Idaho Falls, Idaho, United States

*Address all correspondence to: nicolas.woolstenhulme@inl.gov

IntechOpen

References

[1] TREAT History. Available from: https://transient.inl.gov/SitePages/TREAT%20History.aspx [Accessed: September 25, 2021]

[2] Mcfarlane DR, Freund GA, Boland JF. Hazards Summary Report on the Transient Reactor Test Facility (TREAT). ANL-5923. 1958

[3] Freund GA, Iskendarian HP, Okrent D. TREAT a Pulsed Graphite-Moderated Reactor for Kinetics Experiments. In: Proc. 2nd United Nations Int. Conf. on the Peaceful Uses of Atomic Energy; Geneva, Switzerland. Vol. 10. 1958. p. 461

[4] Handwerk JH, Lied RC. The Manufacture of the Graphite-Urania Fuel Matrix for TREAT. ANL-5963. 1960

[5] TREAT Baseline Description Document. ANL/RAS-72-73. Argonne National Laboratory; 1972

[6] Bess JD, Dehart MD. Baseline Assessment of TREAT for Modeling and Analysis Needs. INL/EXT-15-35372. Idaho National Laboratory; 2015

[7] Holschuh T, Woolstenhulme N, Baker B, Bess J, Davis C, Parry J. Transient reactor test facility advanced transient shapes. Nuclear Technology. 2019;**205**(10):1346-1353

[8] Bess JD, Woolstenhulme NE, Davis CB, Dusanter LM, Folsom CP, Parry JR, et al. Narrowing transient testing pulse widths to enhance LWR RIA experiment design in the TREAT facility. Annals of Nuclear Energy. 2019;**124**:548-571

[9] Grund JE, et al. Subassembly Test Program Outline for FY 1969 and 1970, IN-1313, IDO-17277. Idaho Nuclear Corporation; 1969

[10] Spencer WA, Jensen AM, McCardell RK. Capabilities of the power

burst facility. In: Proc. Int. Topl. Mtg. Irradiation Technology; September 28-30, 1982. pp 175-198. Available from: https://link.springer.com/chapter/10.1007%2F978-94-009-7115-8_13

[11] Modro SM, et al. Review of LOFT Large Break Experiments. NUREG/IA-0028. U.S. Nuclear Regulatory Commission; 1989

[12] Wright AE, et al. Mark-III integral sodium loop for LMFBR safety experiments in treat. In: Proc. Conf. on Fast, Thermal, and Fusion Reactor Experiments; Salt Lake City, Utah: American Nuclear Society; April 12-15, 1982. Vol. 1. 1982

[13] Woolstenhulme N, Baker C, Jensen C, Chapman D, Imholte D, Oldham N, et al. Development of irradiation test devices for transient testing. Nuclear Technology. 2019;**205**(10):1251-1265

[14] Crawford DC, Swanson RW. RIA testing capability of the transient reactor test facility (IAEA-TECDOC--1122). International Atomic Energy Agency (IAEA). 1999. Available from: https://inis.iaea.org/search/search.aspx?orig_q=RN:30057719

[15] Wachs DM. Transient testing of nuclear fuels and materials in the United States. Journal of the Minerals Metals and Materials Society. 2012;**64**(12):1396

[16] Special issue on restarting the transient reactor test facility. Nuclear Technology. 2019;**205**:10

[17] Gerstner DM, Parry JR, Broussard DJ, Moon BL, Laporta AW, Forshee CP, et al. Safety strategy and update of the TREAT facility safety basis. Nuclear Technology. 2019;**205**(10):1266-1289

[18] Laporta AW. Transient reactor test (TREAT) facility initial approach to

restart criticality following extended standby operation. Nuclear Technology. 2019;**205**(10):1290-1301

[19] Heath BK, Cody CRACE. TREAT restart project. Nuclear Technology. 2019;**205**(10):1369-1377

[20] Thompson SJ, Johnson JT, Schley RS, Townsend CH, David LC. Preliminary modeling to support the TREAT hodoscope system for fuel motion monitoring. In: Proc. Int. Conf. PHYSOR 2016; Sun Valley, ID; American Nuclear Society; May 1-5, 2016. Available from: https://inldigitallibrary.inl.gov/sites/sti/sti/Sort_8873.pdf

[21] Bess JD, Woolstenhulme NE, Jensen CB, Parry JR, Hill CM. Nuclear characterization of a general-purpose instrumentation and materials testing location in TREAT. Annals of Nuclear Energy. 2019;**124**:270-294

[22] Woolstenhulme N, Fleming A, Holschuh T, Jensen C, Kamerman D, Wachs D. Core-to-specimen energy coupling results of the first modern fueled experiments in TREAT. Annals of Nuclear Energy. 2020;**140**:107117

[23] Schulthess J, Woolstenhulme N, Craft A, Kane J, Boulton N, Chuirazzi W, et al. Non-Destructive post-irradiation examination results of the first modern fueled experiments in TREAT. Journal of Nuclear Materials. 2020;**541**:152442

[24] Kamerman D, Woolstenhulme N, Imholte D, Fleming A, Jensen C, Folsom C, et al. Transient testing of uranium silicide fuel in zircaloy and silicon carbide cladding. Annals of Nuclear Energy. 2021;**160**:108410

[25] Armstrong RJ, Folsom CP, Fleming AD, Jensen CB. Results of the CHF-SERTTA in-pile transient boiling experiments at TREAT. In: Proceedings of TopFuel 2021; Santander, Spain: American Nuclear Society; 10/24/2021-10/28/2021. INL/CON-21-64083. 2021.

Available from: https://www.osti.gov/servlets/purl/1820615

[26] Woolstenhulme N, Jensen C, Folsom C, Armstrong R, Yoo J, Wachs D. Thermal-hydraulic and engineering evaluations of new LOCA testing methods in TREAT. Nuclear Technology. 2021;**207**(5):637-652

Chapter 3

Experimental Breeder Reactor II

Chad L. Pope, Ryan Stewart and Edward Lum

Abstract

The Experimental Breeder Reactor II (EBR-II) operated from 1964 to 1994. EBR-II was a sodium-cooled fast reactor operating at 69 MWth producing 19 MWe. Rather than using a loop approach for the coolant, EBR-II used a pool arrangement where the reactor core, primary coolant piping, and primary reactor coolant pumps were contained within the pool of sodium. Also contained within the pool was a heat exchanger where primary coolant, which is radioactive, transferred heat to secondary, nonradioactive, sodium. The nuclear power plant included a sodium boiler building where heat from the secondary sodium generated superheated steam, which was delivered to a turbine/generator for electricity production. EBR-II fuel was metallic uranium alloyed with various metals providing significant performance and safety enhancements over oxide fuel. The most significant EBR-II experiments occurred in April 1986. Relying on inherent physical properties of the reactor, two experiments were performed subjecting the reactor to loss of primary coolant flow without reactor SCRAM and loss of the secondary system heat removal without reactor SCRAM. In both experiments, the reactor experienced no damage. This chapter provides a description of the most important design features of EBR-II along with a summary of the landmark reactor safety experiments.

Keywords: fast reactor, sodium-cooled reactor, metal fuel, inherent safety, breeder reactor

1. Introduction

The worldwide nuclear power industry is currently dominated by light water reactor technology. However, U-235 fissile material resource utilization challenges are likely to drive the need for non-light water reactor technologies when one considers timelines extending beyond the next half century. Many alternative reactor technologies that are capable of addressing the resource constraints of light water reactors are currently being pursued.

It is frequently worthwhile to look to the past as a means of guiding the path for the future. The first demonstration nuclear power plant was influenced by the expectation that limited supplies of fissile material will necessitate breeding fissile material. The Experimental Breeder Reactor I (EBR-I) achieved initial power production operation on December 20, 1951 (see **Figure 1**) and produced the first significant amounts of electrical energy generated by nuclear fission. EBR-I was a sodium-potassium cooled fast neutron spectrum reactor capable of breeding more fissile material than it consumed. The reactor was part of a power plant design that included steam generation and a turbine/generator system.

Following the success of EBR-I, the Experimental Breeder Reactor II (EBR-II) was constructed near EBR-I on the high-altitude arid Snake River Plain of

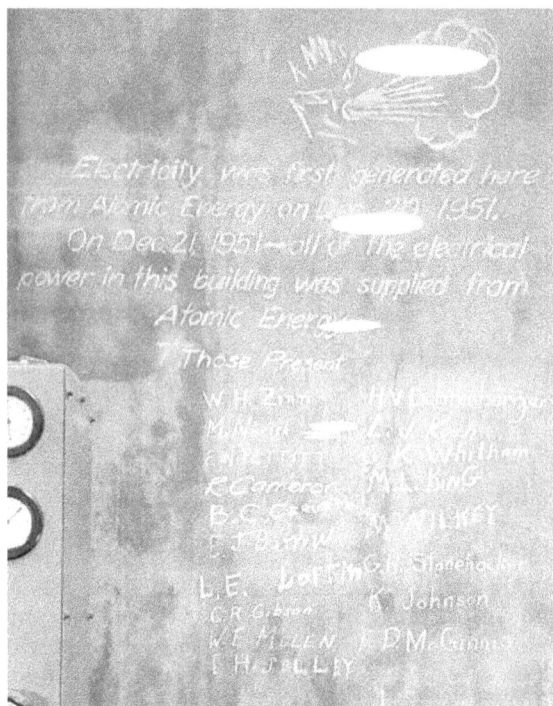

Figure 1.
Chalk message at EBR-I [1].

southeastern Idaho in the western United States. Like EBR-I, EBR-II was a complete power plant demonstration, and it also included an attached fuel cycle facility to reprocess spent fuel using a melt refining process (see **Figure 2**). The reactor was a sodium cooled fast reactor (SFR) capable of producing more fissile material than it consumed. EBR-II achieved initial criticality in 1964 and operated until 1994. The reactor produced 19 MWe and supported decades of sodium cooled fast reactor development activities. The success of EBR-II provides insight into the potential benefit of future widespread use of sodium cooled fast reactors as a means of addressing fissile material resource limitation issues. It should also be noted that numerous other sodium cooled fast reactors have been developed including, but not limited to, Fermi I and the Fast Flux Test Facility in the US, Phénix and Super Phénix in France, Joyo and Monju in Japan, BN-350 in Kazakhstan, BN-600 and BN-800 in Russia, as well as sodium cooled fast reactors in India and China.

From an industry perspective, there is a resurgence of interest into sodium cooled fast reactors. Two commercial entities have proposed the use of sodium cooled fast reactors. The TerraPower company is pursuing a sodium cooled fast reactor coupled with a molten salt heat storage capability. The reactor is capable of producing 345 MWe as well as boosting the output to 500 MWe by using heat stored in molten salt. The reactor is called Natrium, which is Latin for sodium. In October 2020, the US Department of Energy awarded TerraPower funding to demonstrate the Natrium technology. TerraPower is targeting 2023 for submission of a construction permit from the US Nuclear Regulatory Commission. The planned location for the reactor will be one of four prospective sites in the state of Wyoming in the western United States. Furthermore, the Oklo Power Company has a sodium cooled fast reactor design which produces 4 MWth and integrates significant inherent safety

Figure 2.
Experimental breeder reactor II [2].

features into the design. Oklo Power submitted the first-ever combined construction and operation license application to the US Nuclear Regulatory Commission in March 2020.

From a US Government perspective, the US Department of Energy is pursuing the Versatile Test Reactor (VTR). The VTR is a sodium cooled fast reactor that will operate at 300 MWth. The purpose of the VTR is to provide a very high neutron flux (4×10^{15} n/cm^2 sec) which will be used to test fuels and components for a wide range of advanced reactor concepts. The VTR project received Critical Decision–1 approval in September of 2020, allowing the project to proceed to preliminary design.

With this information in mind, it is worthwhile to reflect on the design and performance EBR-II since it provides tremendous knowledge and potential direction for sodium cooled fast reactors moving forward.

2. Power plant and reactor design

EBR-II was a complete power plant along with an attached fuel cycle facility. The reactor containment was centered between the sodium boiler building and the turbine/generator building. The reactor was an SFR which acted as a breeding facility and test bed for liquid metal fast breeder reactors [3]. Along with this, EBR-II produced electricity as part of its overall demonstration. Being a fast neutron spectrum reactor, the neutron chain reaction was driven primarily by fast neutrons. Fast neutrons often invalidate many assumptions commonly assumed for light water reactors. The long neutron mean free path associated with a fast neutron spectrum is indicative that much of the core is coupled, meaning there are relatively few localized reactivity effects. This often helps prevent localized peaking. The long mean free path of neutrons also means that negative reactivity insertion due to control rods in a few sections of the core provides the necessary means to shut down the reactor.

EBR-II was a pool-type SFR, meaning the core, and all supporting structures, were contained in a double walled vessel comprised of 86,000 gallons of primary sodium [4]. Due to this design, leaks in any of the primary system piping would drain into the primary coolant. This would result in a loss of plant efficiency but would not leak primary sodium outside the vessel. This design is unlike loop type reactors (i.e. Fast Flux Test Facility, Monju, SuperPhenix), where a leak in the primary coolant had the potential to cause a sodium fire and release activated sodium and would likely cause prolonged outages for repairs.

From a reactor operating perspective, sodium couples four very important properties: 1) extremely high boiling point (870 C) at atmospheric pressure, 2) outstanding heat transfer properties owing to its metallic nature, 3) relatively high atomic weight compared to neutrons leading to limited neutron modera-tion, and 4) a low neutron absorption cross section along with a relatively short neutron activation half-life of 15 hrs. These properties allow sodium to be used as an outstanding fast reactor coolant. The most obvious drawback of using sodium metal as a reactor coolant is the fact that it reacts with water and evolves hydrogen in the reaction process. The sodium-water reaction can be violent especially when the evolved hydrogen combines with oxygen. The reaction between sodium and water follows two primary schemes forming sodium hydroxide and sodium oxide as shown in Eqs. (1) and (2). In both reactions, hydrogen is also produced which presents a flammability and explosion hazard. It is important to keep in mind that a leak of high temperature sodium to an air atmosphere will result in dense white smoke which makes leak identification simple.

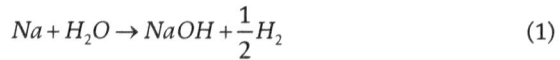

$$Na + H_2O \rightarrow NaOH + \frac{1}{2}H_2 \tag{1}$$

$$2Na + H_2O \rightarrow Na_2O + H_2 \tag{2}$$

The primary coolant arrangement for EBR-II can be seen in **Figure 3**. This highlights the major components associated with the primary coolant. Cold coolant (~370 C) was drawn in via two primary pumps, each of which supplied ~18,000 liters per minute of coolant and was split into a high-pressure and lower pressure inlet plenum at the bottom of the core. Of special note, the two primary coolant pumps were single-stage centrifugal mechanical pumps: a first of their kind for liq-uid metal coolant at the time. After flowing through the core, hot coolant (~480 C) then flowed into a shared upper plenum with a single outlet (shown as a "Z" in both figures). The hot coolant then entered the heat exchanger and was discharged back into the primary coolant pool. To filter out impurities, a cold-trap system continually filtered primary coolant by reducing the sodium temperature to reduce the solubil-ity limits and precipitate out impurities. Above the sodium was ~12 in. of argon gas providing a protective inert cover for the sodium coolant.

The secondary system extracted heat from the primary system which was then used to drive a Rankine cycle for power generation [4]. The sodium flow rate for the secondary system was 23,000 liters per minute, with an inlet temperature of 310 C and an outlet temperature of 460 C. Transferring heat from the radioactive primary sodium to non-radioactive secondary sodium provided a safety enhancement and the ability to place much of the secondary system in a separate sodium boiler build-ing, which was physically separate from the main reactor building. This separation reduced the time required in containment and reduced the potential for radioactive impurities to cause exposure. The sodium boiler building design incorporated a

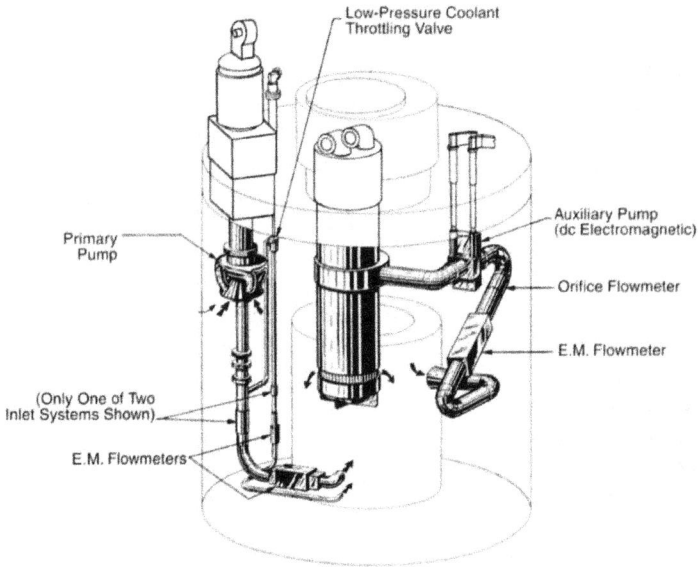

Figure 3.
Primary coolant system for EBR-II [3].

sacrificial plastic wall located away from the reactor building. The sacrificial wall would fail in the event of a catastrophic sodium water reaction in the sodium boiler building thereby directing the reaction energy away from the reactor building.

For the Rankine cycle, superheated steam was generated at 450 C with a pressure of 9000 kPa: this powered an off-the-shelf 20 MW turbine generator. The ability to use off the shelf components, helped reduce cost in the secondary system (one of the primary objectives of EBR-II). The secondary system allowed for a steam by-pass to continually dump heat despite any electrical needs. The overall EBR-II heat transfer pathway is shown in **Figure 4**.

In addition to the primary and secondary systems, an auxiliary pump was used to ensure a low-pressure flow rate was always present, despite normal power failure.

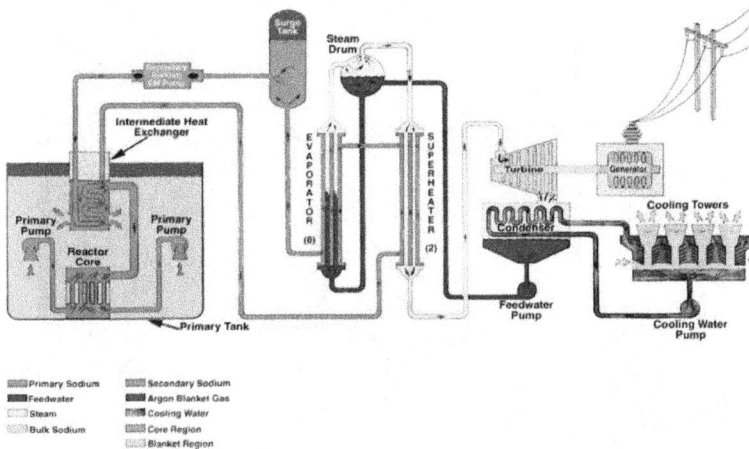

Figure 4.
EBR-II heat transfer pathway [2].

The auxiliary pump was attached to a DC battery system, which would last long enough to allow the EBR-II system time to convert from forced cooling to natural circulation. To aid in the natural circulation, two shutdown coolers penetrated the primary coolant tank and allowed for heat removal directly to the atmosphere. The shutdown coolers contained sodium-potassium which extracted heat from the primary system and was exposed to an air-cooled heat exchanger.

The EBR-II core used 637 hexagonal subassemblies that made up the driver, inner blanket, and outer blanket regions. **Figure 5** shows the top of the reactor core prior to the introduction of sodium coolant. The driver region was where a majority of the neutron flux was generated, which meant that a majority of the power was generated in this region. In terms of an equivalent cylinder, EBR-II had a diameter of ~20 in. and a height of ~14 in. Subassemblies were generally broken up into a few major categories: driver, blanket, control, reflector, and experiment [5].

Subassembly types shared many characteristics, the most notable being the outer dimensions which allowed for subassemblies to be moved throughout the core, depending on the specific needs. Each assembly was hexagonal in shape, and had an outside flat-to-flat distance of 5.82 cm with a flow duct wall thickness of 0.10 cm. All subassemblies also had an upper adapter (this allowed for subassemblies to be placed and removed from the core), and a lower adapter. The lower adapters had slightly different configurations to ensure subassemblies were placed in the correct location.

Driver fuel assemblies were comprised of, in general, a lower adapter, fuel pin grid, and upper preassembly. Coolant flowed from the inlet plenum into the lower adapter, through fuel pin grid (where heat was transferred to the coolant), and out the upper preassembly into the outlet plenum. Multiple driver fuel designs were used throughout the lifetime of EBR-II, and as such, a brief description of the MK-II fuel assembly design is given [5]. Since these were used throughout the life of the reactor. Comprised within the fuel pin grid were 91 fuel pins in a hexagonal lattice with a fuel pitch of 0.56 cm. Fuel pins are described further in a later section. Half-worth driver assemblies where nearly identical to driver fuel assemblies, however, half of the fuel pins were replaced with stainless steel pins; this reduced the reactivity of the fuel assembly. Half-worth driver assemblies were typically placed near the center of the core to dampen peaking effects.

Figure 5.
EBR-II reactor Core [4].

Blanket assemblies were used throughout the life of EBR-II, where they were initially inserted around the core to breed plutonium. Blanket assemblies contained 19 fuel pins comprised of a fuel slug (outer diameter (OD) 1.1 cm), sodium bond (OD 1.16 cm), and a stainless-steel cladding (OD 1.25 cm). Blanket fuel pins were much larger than their driver counterparts due to the lower power density and a desire to increase the fuel to sodium ratio to promote breeding in the pins. Blanket fuel pins were 1.43 m long.

EBR-II, like many SFRs, used full assembly positions for the safety and control rods (denoted control assemblies from here on). Control assemblies had an inner hexagonal duct (flat-to-flat diameter of 4.90 cm) which contained a fuel region with 61 fuel pins which could be brought into the plane of the driver fuel to add reactivity to the core. Some control rods (designated high worth control rods) had a region comprised of seven B_4C pins directly above the fuel, which acted as an additional poison to ensure the reactor could shut down and remain shut down.

Reflector assemblies did not contain a pin grid section, but instead contained stacks of stainless-steel hexagonal blocks. These blocks were used to reflect neutrons back into the core and were typically placed in the periphery.

Experimental assemblies were unique in both design and contents. These assemblies maintained the hexagonal duct but could contain fuel, material, monitor, etc. experiments. Experimental assemblies are described in greater detail in a subsequent section.

Fuel pins consisted of a metallic fuel slug (OD of 0.33 cm), sodium bond (OD 0.38 cm) and stainless-steel cladding (OD 0.44 cm). The total length of the fuel pin was 62.04 cm, where the fuel slug had a length of 34.29 cm. Above the fuel slug was a helium plenum to capture fission product gasses and was often tagged with trace amounts of xenon to allow for the determination of burst fuel pins. Each fuel pin was surrounded by a wire-wrap with a diameter of 0.125 cm and an axial pitch of 15.24 cm. The wire wrap was used to ensure fuel pins did not come in contact with each other and provided additional coolant mixing to encourage heat transfer. Throughout the lifetime of EBR-II, the fuel pins changed slightly in dimensions, however, the dimensions presented provide a reasonable representation of a typical fuel slug. **Figure 6** shows an arrangement of driver fuel pins along with the wire-wrap.

The fuel slugs in Mk-II subassemblies comprised a uranium-fissium alloy (95 wt. % uranium 5 wt. % fissium), meaning that the fuel was metallic in nature, compared with the typical ceramic fuel (uranium-oxide) found in light water reactors. The uranium in the fuel was enriched to between 45 wt. % and 67 wt. % U-235, again in stark contrast to the typical 5 wt. % light water reactor fuel. Fissium was comprised of elements to simulate dominant mid-fuel cycle fission products. The short-highly-enriched fuel for EBR-II created a very short-flat core, which provided multiple inherent safety benefits, described in greater detail later.

One other noteworthy feature of the EBR-II design involved a fuel storage basket located within the primary tank. The fuel storage basket contains 75 indexed storage tubes in three concentric rings. Each tube could accommodate a single fuel assembly. The fuel storage basket was accessed essentially anytime by operators including when the reactor is operating at full power. The fuel storage basket provided great operational flexibility. During reactor operation, spent fuel assemblies stored in the basket could be removed one at a time and transferred out of the reactor facility and delivered to a hot cell facility for storage and disassembly. Fresh fuel and experimental assemblies could also be loaded into the basket during reactor operation. When the reactor was shut down, operators could then quickly move spent fuel assemblies from the core into the fuel storage basket and move fresh fuel from the basket into the core making the refueling outage time as short as possible. Since the driver

Figure 6.
Fuel pin arrangement [6].

region of the core contained roughly 100 assemblies, the 75-assembly fuel storage basket provided ample capacity for staging fresh fuel assemblies as well as holding spent fuel assemblies removed from the core. With the fuel storage basket located within the primary tank, the sodium coolant provides sufficient heat transfer capacity to ensure the spent fuel assemblies are adequately cooled prior to their removal.

3. Experiments

Experiments were not placed in specific assembly locations in the core. This is unlike many light water test reactors which have specific ports or testing locations. Instead, experiments were often placed in the same hexagonal duct as a typical driver fuel assembly. This meant multiple experiments could be placed in the same assembly, experiments could be intermixed with fuel pins, or experiments could be placed in an assembly with dummy stainless-steel pins. The placement of an experiment in the core was largely determined by the conditions required for the experiments. If an experiment needed a large flux of high energy neutrons in a short period of time, it could be placed in the center of the core. On the other hand, if an experiment needed to experience a large neutron fluence over a long period of time, it could be placed in the periphery of the core. Overall, an experiment could likely be placed in any assembly position within the core, with the exception of the control/safety assemblies. To compensate for any loss of reactivity due to adding experimental assemblies, additional driver assemblies were placed in the periphery of the core.

EBR-II also examined multiple endurance type testing for both fuel and cladding [7]. In the 1970's, a series of experiments examined running fuels to cladding breach (RTCB) and running fuel beyond cladding breach (RBCB). These experiments were used to help increase the burnup capabilities for fuels and determine neutron fluence limits for these fuels. To accomplish this, an additional cover-gas cleanup system (GGCS) was installed to help remove radioisotopes that leaked from the fuel and into the argon cover gas.

3.1 Dry/wet critical experiment

In April of 1961, before EBR-II was used as a power producing or breeding facility, it underwent a series of zero power experiments (designated as less than 1 kW of power) before the primary system was filled with sodium [8–10]. To perform the dry critical experiment, fuel and blanket assemblies that would be used for normal operations were loaded into the core in a similar configuration to when sodium would be added. For this, additional driver assemblies (~87 driver assemblies compared with ~56 driver assemblies for a sodium filled core) were required achieve criticality since the lack of sodium increased neutron leakage in the core. These experiments were able to take place while construction work was being performed elsewhere in the plant.

The basis of these experiments was twofold. The first was used to determine the performance of the system without sodium, which allowed them to subsequently identify sodium effects on system neutronics. The second gathered operational data to determine if modifications or improvements were required prior to adding sodium. To gather this information, four major experiments were conducted. The first was to determine the strength of the neutron source and the neutron detector responses to ensure an adequate relationship between the two. The second was an approach to critical to verify the ability to insert assemblies and determine the dry critical mass. The dry critical mass could then be compared with the wet critical mass to determine the total reactivity worth of the sodium. The third aspect examined the neutron flux distribution and fission distribution throughout the core and provided a power calibration. The final aspect that was examined was a series of reactivity measurements. This included seven measurements ranging from the total worth of the control rods, individual control rods, to the dry isothermal temperature coefficient of reactivity.

3.2 Connected fuel cycle

EBR-II was originally designed as a power-producing facility which would be able to produce more fuel (in the form of plutonium) than it consumed. To accomplish this, blanket subassemblies were placed around the periphery of the core, where neutrons which leaked out would be absorbed by U-238 to produce plutonium. In addition to creating a core design which was favorable for generating fuel, additional facilities were constructed on-site to allow for fuel/experiment examination and fuel reprocessing.

The fuel cycle facility (FCF) was built to allow for post-irradiation examination of experiments placed in the core [11]. FCF allowed for experiments to be removed from one subassembly and placed in a new subassembly for further irradiation if necessary. Along with this, FCF was used to reprocess spent EBR-II fuel using a crude melt refining technique rather than a complicated and large solvent extraction process. Melt refining involved melting the spent fuel elements and mechanically separating fission products and slag from the uranium. The uranium (or other actinides) was then used to fabricate additional fuel.

The last decade of operations for EBR-II was focused on the Integral Fast Reactor (IFR) concept [12, 13]. This project encompassed nearly all aspects of life for a nuclear reactor. The IFR concept was meant to overcome many obstacles such as proliferation concerns, waste generation concerns, and reactor safety concerns. The IFR concept was meant to provide the United States (and the world) with a nuclear energy concept that could provide a nearly inexhaustible energy supply for the future. Unfortunately, in 1994, the IFR concept and indeed EBR-II was terminated, and the full realization of the IFR concept never came to pass.

3.3 High burnup

One of the many advantages of fast reactor technology is the ability to "burn" to a greater extent than thermal reactor. The average burnup for a typical light water reactor is 45,000 MWD/MTHM. EBR-II demonstrated 20 atom % burnup which is the equivalent of 190,000 MWD/MTHM. These burnups are possible primarily because of the fast neutron spectrum present in the reactor. Along with the energy extracted from the fission of U-235, the fast spectrum transmutes the U-238 to higher order actinides. Those elements are subsequently fissioned, releasing energy rather than creating a problematic waste issue. The transmutation process does happen in thermal spectrum reactors, but to a far lesser extent. Given this, the extractable energy from fast reactors is fundamentally limited by the structural materials of the fuel and how long they can serve the engineering requirements under significant irradiation.

3.4 Inherent safety

April 3rd 1986 is a date that is unknown to the general public and to large portions of the nuclear industry. The reason was that nothing newsworthy happened that day. The EBR-II functioned as designed without any damage, everyone working in the facility went home that day, and in general it was like any other day in southeast Idaho. Despite nothing being widely reported that day, one of the most significant achievements in nuclear reactor technology was demonstrated. The EBR-II was intentionally placed into an accident scenario that would have melted down any light water reactor. The accident scenario far exceeded that of Three Mile Island. The scenario was to operate the EBR-II at 100% power, disable the primary coolant pumps (for the first experiment) and the secondary cooling pumps (for the second experiment). Both experiments were conducted without SCRAM the reactor. To achieve the plant conditions listed above, EBR-II was modified to create the conditions but still remain in control in case unpredictable behavior occurred. An example of a modification was the cooling pumps. They were not directly disabled; the pump controllers were modified to simulate coast down function shapes, one of which simulated station blackout. Nominally the presented scenario would be a guaranteed melt-down for the typical US nuclear power plant. The EBR-II design, however, managed to achieve a temperature profile shown in **Figure 7**.

Figure 7 demonstrates that given a catastrophic failure of major safety mechanisms, including failure to SCRAM following the loss of primary reactor coolant pumps or secondary coolant pumps, the peak temperature remained well below the sodium coolant boiling temperature of 870 C. Additionally, the peak temperature only lasted tens of seconds before reducing to a temperature less than that of 100% power. The inherent properties of the reactor design drove the reactor response rather than any engineered active systems. In short, the large thermal mass of the primary coolant pool, the thermal expansion of the core upon heating and the properties of the metal fuel all worked together to cause the reactor to become subcritical before fuel damage occurred following termination of coolant pump operation even without reactor SCRAM. The current fleet of light water reactors subjected

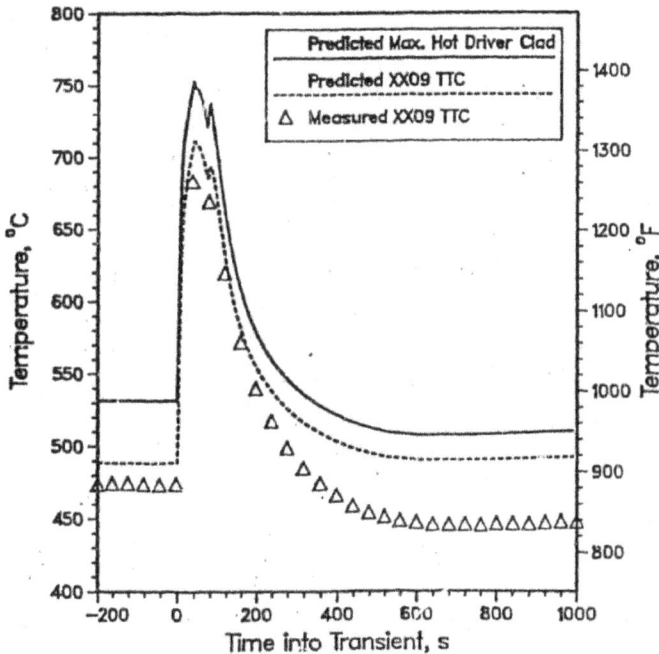

Figure 7.
EBR-II driver temperature predicted and measured [14].

to a similar experiment would melt down without active cooling because the water coolant would eventually boil and the heat removal would be insufficient to prevent fuel melting.

Removal of the heat from the fuel elements and transporting that heat to the outside required several design layers. The first layer starts with the fuel elements, the metallic uranium, sodium bond, and stainless steel 316 cladding which provides an uninterrupted metallic conduction path from the uranium slugs to the sodium coolant. Sodium has one thousand times the heat conduction of water and in EBR-II's design, allowed for the decay heat to be transported rapidly to the sodium pool. **Figure 8** shows the uninterrupted metallic conduction path, the sodium is the green color.

The second layer was the large sodium pool that could absorb a significant amount of heat without changing temperature. Even without active cooling, the natural convection of the sodium over the fuel elements was enough to circulate cool sodium in from the pool and inject hot sodium back in the pool. Given the 337,000 liters of sodium in the pool, it would take many weeks for the pool to reach a temperature where the sodium would begin to boil.

The last layer was the natural convection heat exchanger that led pool sodium to a chimney that naturally exhausted to the outside. The heat exchanger functioned solely on the temperature differential of the pool to the outside and required no external power. The natural convection heat exchanged moderated the temperature in the pool to keep the sodium from boiling away.

In summary, the solution to a run-away heating event was to increase the thermal conduction from the fuel slugs to the outside to the point where the heat generated could not exceed the bandwidth of the heat removed to the outside.

The previous sections describe how EBR-II removed the decay heat from the fuel elements, mitigating a meltdown event. This mitigation only covered long term

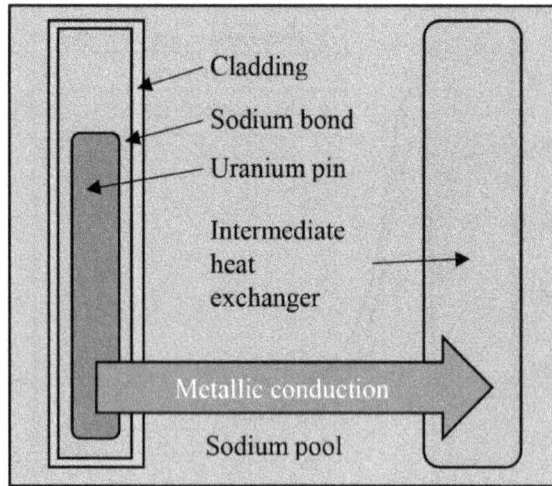

Figure 8.
Thermal conduction path [15].

inherent safety, not short term. Short term transients also require mitigation due to their rapid onset. Large reactivity insertions can cause localized heating that cannot be conducted away fast enough leading to fuel melting. An example is, during fuel shuffle operations, an assembly falls into the pool. Mitigation of these events (aside from not causing them in the first place) requires a negative feedback mechanism to compensate for the reactivity change. In reactors, these are called negative reactivity coefficients. They are a result of the inherent physics of a reactor's design and are nominally passive. For example, as a legal requirement in the US, light water reactors have a negative temperature coefficient. Meaning, the hotter the fuel, the less fission occurs, thus preventing a cascade event where heating creates more fission which creates more heating. For EBR-II several of these coefficients kept the reactor in a 100% negative feedback regime.

The first of these and most effective was the expansion of the sodium inside of the core region. The liquid sodium density reduced due to thermal expansion. Given that sodium has a moderating effect on fast neutrons, the decrease in moderation led to an overall negative reactivity feedback due to sodium temperature increases. This proved invaluable in the safety heat removal tests because as the temperature increased, there was a greater the reduction in fissions.

Second, EBR-II's core construction allowed for thermal expansion in the core. As temperature increased the fuel assemblies were pushed away from each other. The core grid plate that locked the bottom of the assemblies would expand due to temperate having the effect of increasing the pitch. Fast reactors in general are very sensitive to geometry changes due to their high-power densities. Any expansion increases the leakage of neutrons due to the increase in effective surface area with the same neutron population.

These two negative reactivities constitute 99% of the reactivity coefficients. They kept the reactor from running away in a thermal transient allowing for thermal conduction to occur. The long-term conductive mechanisms of EBR-II then kept the reactor from melting down. With these passive mechanisms in place, the severe accident scenario described in the previous section could happen without any real consequences. It was due to the inherent safety mechanisms of EBR-II that made April 3rd 1986 just another day in southeast Idaho.

4. Conclusion

EBR-II was arguably the most significant, meaningful, and successful sodium cooled fast reactor power plant demonstration in the history of nuclear power. It must be emphatically stated that the success of EBR-II was the result of actual demonstration rather than simulation and modeling or claims of future success based on short-lived small past experiments. Over a 30-year operating lifetime, the reactor demonstrated all aspects necessary for using a sodium cooled fast reactor for power production. Numerous technological advancements were made using EBR-II. Foremost among the advancements were 1) the demonstration of a pool type primary coolant arrangement with all primary piping and pumps located within the pool, 2) the ability to conduct fuel handling activities in opaque molten sodium, 3) the ability to transfer fuel into and out of the primary sodium pool while the reactor was operating at full power, 4) the ability to safely operate a system where heat is transferred from molten sodium to water, 5) the development of metallic fuel, 6) the demonstration of tremendous fuel burnup, and 7) the demonstration of compact on-site fuel reprocessing. The most significant accomplishment of EBR-II was the demonstration of the inherent safety associated with the overall reactor design and material properties that allowed the reactor to survive the most severe accident scenarios, loss of flow without SCRAM and loss of heat sink without SCRAM, with no fuel damage.

It is hoped that the success of EBR-II will not only influence the design of future sodium cooled fast reactors, but that it will be identified as an example of the true feasibility of such designs. This chapter is dedicated to the memory of Len Koch who was present for the startup of EBR-I and served as one of the principal EBR-II designers.

Author details

Chad L. Pope[1]*, Ryan Stewart[2] and Edward Lum[3]

1 Idaho State University, Pocatello, Idaho, USA

2 Idaho National Laboratory, Idaho Falls Idaho, USA

3 Los Alamos National Laboratory, Los Alamos, New Mexico, USA

*Address all correspondence to: popechad@isu.edu

IntechOpen

References

[1] Pope C. Photo of EBR-I Turbine Area, Arco, ID; 2011

[2] EBR II Fuels and Irradiation Physics Database. Argonne National Laboratory. https://fipd.ne.anl.gov/about/. [Accessed: June 9, 2022]

[3] Sackett J. Operating and Test Experience of EBR-II, ANL/CP-72616. Idaho Falls, ID: Argonne National Laboratory; 1991

[4] Koch L. Experimental Breeder Reactor-II (EBR-II) an Integrated Experimental Fast Reactor Nuclear Power Station. Lemont, IL: Argonne National Laboratory; 2008

[5] Lum E et al. Evaluation of Run 138B at Experimental Breeder Reactor II, a Prototypic Liquid Metal Fast Breeder Reactor. OECD: International Handbook of Evaluated Reactor Physics Benchmark Experiments; 2018;1-217

[6] Pope C. Photo of EBR-II Fuel Pin Grid, Pocatello, ID; 2020

[7] Perry W et al. Seventeen years of LMFBR experience: Experimental breeder reactor II (EBR-II). In: American Power Conference. Lemont, IL:Argonne National Laboratory; 1982

[8] Koch L et al. EBR-II Dry Critical Experiments Experimental Program, Experimental Procedures, and Safety Considerations, ANL-6299. Lemont, IL:Argonne National Laboratory; 1961

[9] McVean R et al. EBR-II Dry Critical Experiments, ANL-6462. Lemont, IL: Argonne National Laboratory; 1962

[10] Kirn F, Loewenstein W. EBR-II Wet Critical Experiments, ANL-6864. Lemont, IL: Argonne National Laboratory; 1964

[11] Stevenson C. The EBR-II Fuel Cycle Story. United States: La Grange Park, IL. American Nuclear Society; N. p; 1987

[12] Till C, Chang Y. Plentiful Energy the Story of the Integral Fast Reactor. CreateSpace; American Nuclear Society. Scotts Vally, CA. 2011

[13] Hannum W. The Technology of the Integral Fast Reactor and its associated fuel cycle. Vally, CA; Progress in Nuclear Energy. 1997;**31**:1-217

[14] Herzog J et al. Code Validation with EBR-II Test Data, ANL/CP-74826. Lemont, IL: Argonne National Laboratory; 1992

[15] Lum E. Graphic Showing Thermal Conduction Path. Pocatello, ID; 2020

Chapter 4

Idaho State University AGN-201 Low Power Teaching Reactor: An Overlooked Gem

Chad L. Pope and William Phoenix

Abstract

A category of reactors called university research and teaching reactors, includes relatively high-power pool-type and low-power solid-core reactors. Many high-power university reactors are largely used for irradiations and isotope production. Their almost constant operation tends to impede student access. A university reactor can be particularly relevant to the university's mission of preparing well-rounded students who have theoretical knowledge, reinforced by focused laboratory reactor experience. The solid-core Idaho State University Aerojet General Nucleonics (AGN) model 201 reactor operates at such a low power (5 W maximum) that it is not useful for isotope production activities. However, the AGN-201 reactor is well suited for teaching and research activities. The solid-core AGN-201 reactor requires no active cooling system, uses a simple shielding arrangement, and the very low operating power results in trivial burnup providing an operating lifetime exceeding many decades. It is thus worthwhile to examine the Idaho State University AGN-201 nuclear reactor more closely because it offers a wide range of research and teaching capabilities while being widely available to students.

Keywords: reactor, solid-core, research reactor, university reactor, low-power reactor

1. Introduction

University research and teaching reactors are fundamentally intended to help prepare nuclear engineering and other students for entry into the nuclear workforce. They introduce students to the disciplined, structured environment of operating a reactor licensed by the Nuclear Regulatory Commission (NRC). They also offer students hands-on experience, provide opportunities to demonstrate the operation of reactors and a variety of the traditional applications of reactors such as neutron activation, and introduce them to the application of nuclear instrumentation, applied principles of health physics, and more. They can be useful to a wide range of people beyond university nuclear engineering students, including those from National Laboratories, utilities, regulators, and others.

The Idaho State University Aerojet General Nucleonics (AGN) model 201 nuclear reactor is an example of a very safe, low-power, solid-core reactor designed with students and teaching in mind. It was developed in the late 1950's by AGN to satisfy the need of university nuclear engineering departments for a relatively inexpensive, safe, flexible and available reactor with a long design life. The AGN-201's safety results from, *inter alia*, its 'thermal fuse' that terminates excessive sustained operation at high power, a large negative temperature coefficient ($-0.035\%\Delta k/k$ °C^{-1}), and low available excess reactivity (nominally 0.18% $\Delta k/k$ ($0.24) at 20°C) [1]. These safety features, and other design features, make it an ideal teaching reactor in an environment with rapid turnover of student operators and other personnel.

The small teaching reactors generally preceded the higher-power reactors at universities. As the university's interest moved to the higher-power reactors, reactors like the AGN-201 s fell into disuse and most were decommissioned. Recently however, there has been a renewed interest in the utility of AGN-201 nuclear reactors [2]. In addition to discussing the potential uses for the AGN-201, this chapter includes discussions of the challenges of replacing obsolete components to facilitate continued operation. To address this issue, Idaho State University teamed with members of the community whose expertise in project management, instrumentation and control, licensing and other subjects complemented the university's expertise and resources.

A nuclear reactor is an example of the integrated operation of many systems to support the operation of a nuclear core. Simple reactors can be excellent examples of the integrated operation of the core, nuclear instrument systems, the reactor operator or 'human in the loop' and control rods and their controls. Although the AGN-201 is a simple reactor, it can be used to measure the operation of the core and understand and gain insight into its operation. It is intended to support teaching, training and research in a wide variety of subjects. For example, human/machine interface studies could even be conducted with the operator to test novel display concepts.

The AGN-201 has a variety of attractive design features. The reactor has direct access to the core via the so-called 'glory hole' that runs horizontally through the reactor center. It also has a graphite thermal neutron column at the top of the reactor and beam ports in the radial portion of the graphite reflector. The core is enriched to a nominal 19.5% and given the reactor's low power, the core should essentially never require replacement [1]. The reactor has extremely low background neutron and gamma flux levels that along with the reactor's unusually sensitive nuclear instrument systems, facilitate a wide range of measurements including some that might not be possible in other reactors. For example, it is possible to observe individual chains of fissions when the core is just barely subcritical and flux has been allowed to decay to very low levels thus allowing measurement of the prompt neutron decay constant using Rossi's-α method [3]. Neutron flux near the allowed maximum power level is high enough to usefully activate foils and illustrate reactor physics principles but too low to result in the accumulation of large amounts of fission products.

The AGN-201 provides and supports a number of potential opportunities for demonstrations and tests that complement the theory from the classroom, research and problem solving. A wide range of demonstrations and tests can introduce students to the instrumentation and activities that are conducted by reactor engineers and reactor operators at higher-powered test and research reactors and commercial power reactors. This knowledge can help an instrumentation and control designer or engineer to produce circuits that are more forgiving of noise and to help a technician to differentiate between electronic noise and normal operation of the detector channel and be more successful in reducing noise.

2. Reactor description

The AGN-201 nuclear reactor is a solid-core reactor with no active cooling system. The core is constructed of nine 25.6-cm diameter fuel disks (see **Figure 1**). Four of the disks are 4-cm thick, three of the disks are 2-cm thick and two of the disks are 1-cm thick. Each 4-cm thick fuel disk contains 96 g of ^{235}U, each 2-cm thick fuel disk contains 58 g of ^{235}U, and each 1-ck thick fuel disk contains 29 g of ^{235}U. The overall core height is 24 cm. A graphite reflector surrounds the core both radially and axially. The graphite reflector is 20-cm thick and has a density of 1.75 g/cm^3. The reactor fuel consists of slightly less than 20 wt. % enriched uranium. The uranium is in the form of 15-micron diameter particles of UO$_2$. The UO$_2$ particles are pressed with 100-micron diameter polyethylene particles. The density of ^{235}U in the UO$_2$-polyethelyene fuel is 61 mg/cm^3 and the overall uranium density in the fuel is 305 mg/cm^3. The mass ratio of uranium to polyethylene is 1:3.16. The approximate critical mass of the AGN-201 reactor is 665 g ^{235}U [1].

The reactor uses a total of four control rods; two safety rods, one adjustable coarse rod, and one adjustable fine rod. The control rods are made from the same UO$_2$-polyethyelene fuel as the core. To ensure safety, the fueled control rods enter the core from the bottom (see **Figure 1**) so that gravity, along with compressed springs, ensure rapid removal upon reactor SCRAM. The bottom four fuel disks as well as the lower reflector have holes drilled through them to accommodate the control rods.

In addition to the control rods, the AGN-201 reactor is equipped with a thermal fuse as an ultimate reactor safety shutdown device. The thermal fuse is located just below the core center line (see **Figure 1**). The fuse is similar in construction to the

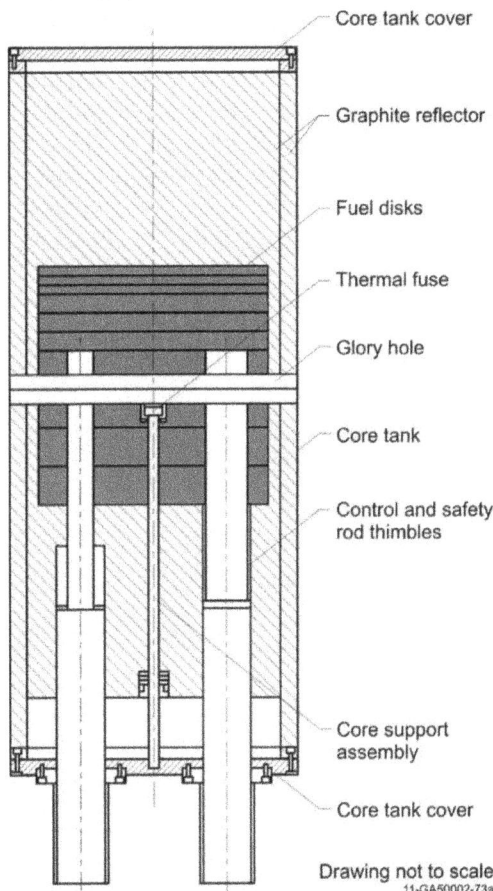

Figure 1.
AGN-201 reactor core, reflector, and control rods [1].

reactor fuel with two key differences. First, rather than polyethylene, the fuse uses polystyrene. Second, the density of uranium in the fuse material is double the value used in the fuel. The two differences coupled with the location of the fuse results in maximizing the fission rate in the fuse compared to all other locations in the core. In the event of a runaway power transient, heat will be generated in the thermal fuse at a greater rate than any other location in the core. As the fuse temperature rises, it will tend to soften when it reaches 100°C and will melt before any the reactor fuel reaches its melting point of 200°C. The AGN-201 reactor design has the lower portion of the core and reflector held in place by the thermal fuse. In the event of the thermal fuse melting, the lower portion of the core and reflector will move downward approximately 5 cm compared to the upper portion of the core which is stationary since it is supported separate from the thermal fuse. The net result will be a dramatic increase in neutron leakage which will terminate the transient. It must be noted that the thermal fuse is a single use safety device.

The reactor core and a portion of the reflector are contained within a gas tight core tank. The core tank is then located within the remainder of the graphite reflector (see **Figure 2**). Surrounding the graphite reflector is a 10-cm thick lead shield. The lead shield is primarily used for gamma ray shielding. The reactor core, reflector, and lead shielding are located within the reactor tank. A graphite thermal column is located on the top lead shielding to support experiments and measurements involving thermalized neutrons. The reactor tank is then located within a 200-cm diameter water filled tank. The radial thickness of the water is approximately 55 cm. The water filled tank is used to absorb neutrons that escape from the core.

To provide access for experiments, a 2.54-cm diameter hole traverses the reactor tank, lead shielding, graphite reflector, and reactor fuel. The hole through the center of the reactor core is commonly referred to as the "glory hole". The glory hole aluminum pipe ensures the core, reflector, lead shield, and water remain properly

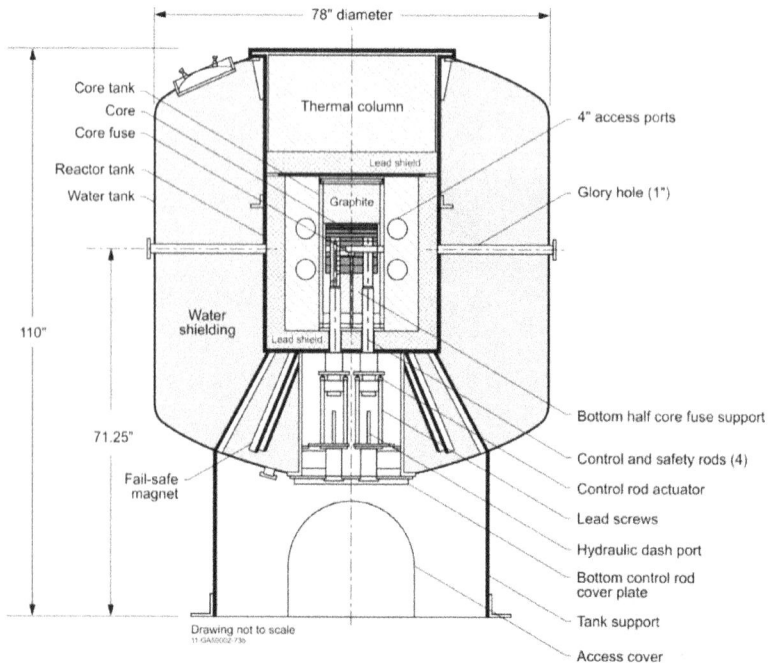

Figure 2.
Reactor tank assembly [1].

sealed. When not in use, the glory hole is typically open to the air atmosphere. When starting the reactor, a neutron source is placed in the glory hole and when the reactor is shut down and not in use, a cadmium neutron absorber is placed in the glory hole to ensure reactor startup cannot occur. In addition to the glory hole, there are four access ports located in the graphite reflector (see **Figure 2**). The access ports are 10.16 cm in diameter and penetrate through the reactor tank, lead shielding and graphite reflector. When not in use, the access ports are typically filled with graphite, lead, and wood (to simulate water).

The reactor radial thermal flux profile is provided in **Figure 3**. The flux profile shows a general Bessel function trend in the core region followed by an exponential drop in the reflector, lead, and water regions. It should be noted that, unlike water reflected thermal reactors, the AGN-201 reactor does not experience an increase in the thermal neutron flux as neutrons enter the reflector. This is primarily due to the difference in neutron scattering properties of water compared to graphite. The flux profile plot demonstrates the effectiveness of the neutron shielding associated with the water shielding tank. The thermal neutron flux at the outer edge of the shielding tank is four orders of magnitude lower than at the center of the reactor.

Reactivity control is carried out using four control rods. Two safety rods, each with a reactivity worth of 1.25% Δk/k ($1.68), are operated in a binary fashion. When starting the reactor, the safety rods are driven fully into the core. No intermediate stopping locations are used for the safety rods. When the reactor is SCRAMed, the safety rods are completely removed form the core. The removal mechanism relies on both gravity as well as compressed springs. A single coarse control rod with

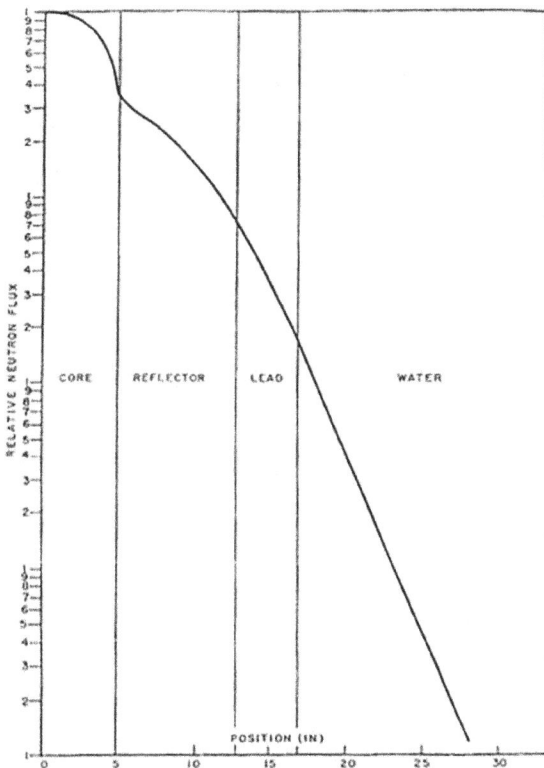

Figure 3.
Reactor radial flux profile [1].

a reactivity worth identical to the safety rods (1.25% $\Delta k/k$ ($1.68)) is raised into the core region during reactor startup. Typically, the coarse control rod is driven to its maximum insertion location, although there are scenarios where the coarse control rod is stopped short of the maximum insertion location. Similar to the safety rods, upon reactor SCRAM, the adjustable coarse control rod is rapidly ejected from the core by relying upon gravity and compressed springs. Finally, the fine control rod has a reactivity worth of 0.31% $\Delta k/k$ ($0.42). The fine control rod is typically driven into the reactor until criticality occurs. Adjustments in the coarse control rod and the fine control rod can then be made to adjust the desired reactor power. Unlike the safety rods and the coarse control rod, the fine control rod is not rapidly ejected from the core when the reactor is SCRAMed. Rather, the fine control rod is driven out of the core at the same rate that it can be driven into the core.

Three monitoring channels are used in the ISU AGN-201 reactor. The three monitoring channel detectors are located within the water filled reactor tank as shown in **Figure 4**. The AGN-201's nuclear instrumentation consists of three different nuclear instrument channels and offer students the opportunity to understand the functions performed by separate portions of the circuit as the incoming signal is processed. Students can study the nuclear instrument channels in a laboratory and then observe them at the reactor.

The three channels are comprised of commercial-grade components. They are more accessible than power plant channels to students and others who wish to study them and their operation over a wide range of neutron flux at an actual reactor. Students can study the instrument systems and their theory and design, and then observe the systems in operation at a wide range of neutron flux. Analog and digital designs of nuclear instrument systems, with a variety of neutron detectors, can be evaluated by using the AGN-201. The AGN-201's nuclear instrumentation consist of the three commonly-found types of nuclear instrument channels that follow the same operating approaches and perform the same functions as the nuclear instrument channels typically found in most reactors. Each channel has a unique but complementary principle of operation. Together, they provide the reactor operators and others with indications of reactor power and the rate of change in power over the entire operating range. Of course they also supply signals for reactor trips.

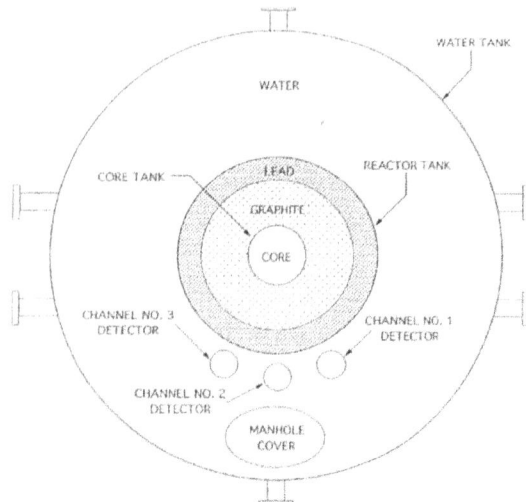

Figure 4.
Reactor assembly plan view [1].

Channel 1 is the startup, source range, channel and uses a BF_3 filled proportional counter. The source range channel illustrates a standard approach that allows the source range channel to display a very low neutron flux in the presence of significant gamma radiation. A proportional-type BF_3 neutron detector produces pulses when gamma radiation and neutrons interact with the BF_3 that fills the detector. The pulses are amplified and shaped, the lower-amplitude pulses due to gamma interactions within the detector are rejected while the remaining higher-amplitude pulses from neutron interactions are further amplified and displayed. The channel displays count rates from the reactor without a source to well above critical. Channel 1 is designed to initiate a SCRAM signal for low power situations when the count rate falls below the setpoint.

Channel 2 is used to monitor the reactor power using a log scale as well as for indication of the reactor period. The channel 2 detector is a BF_3 filled ionization chamber. Channel 2 generates a SCRAM signal when the reactor power falls below 3 x 10^{-13} W or when the reactor power exceeds 5 W. Additionally channel 2 generates a scram signal if the reactor period is less than 5 seconds. The wide range logarithmic neutron instrument channel (channel 2) illustrates a standard approach that allows the channel to detect and display a current signal that is proportional to power over 7 decades. Channels 1 and 2 rely on different applications of wide-range logarithmic amplifiers. The source range nuclear instrument channel's wide-range logarithmic amplifier converts the frequency of incoming pulses from neutron interactions to voltage. The wide range logarithmic current channel's amplifier converts a direct current to a voltage. In both cases, variations in count rate or current level that are due to the normal and expected variations in neutron flux are often misinterpreted as 'noise' that can lead to the period meters having too much variation to be useful indicators to the reactor operators, and the period circuits spuriously tripping. Circuit designers frequently assume the neutron signal is relatively constant and do not anticipate the large noise component that is inherent due to sources. The AGN-201 provides the actual variations in neutron flux that drive oscillations in period meters and indications of reactor power and can be used to evaluate the effect of circuit modifications to reduce the amplitudes of the oscillations.

Channel 3 is used to monitor reactor power using a linear scale. The channel 3 detector is a BF_3 filled ionization chamber. Channel 3 generates a SCRAM signal when the reactor power exceeds 5 W or whenever the linear rotating switch indicator is less than 5% or greater than 95% of full scale.

Figure 5 shows the SCRAM circuit arrangement for the three monitoring channels. It is important to recognize that the SCRAM circuit arrangement is a single signal SCRAM [4]. If any one of the channels identifies a situation that triggers a SCRAM, the reactor will be SCRAMed. That is, the AGN-201 SCRAM circuit is not a two-out-of-three arrangement.

In addition to the monitoring channels, a series of additional interlock circuits are used to prevent reactor startup or to SCRAM the reactor in the event of undesired situations (see **Figure 6**) [4]. The reactor shielding tank water temperature is monitored to ensure that the maximum allowed excess reactivity is not exceeded. If the reactor water temperature falls below 15°C the reactor excess reactivity is unacceptably large and reactor operation is prevented or discontinued. The reactor shielding tank water level is monitored to ensure sufficient shielding is present. Finally, a seismically activated switch is used to prevent reactor operation or discontinue reactor operation in the case of a seismic event. Similar to the reactor monitoring channels, the interlocks follow a series approach so that if any one of the interlocks is triggered, the reactor will not be allowed to operate.

The reactor is operated from a relatively simple console located in the same room as the reactor. The original console was used for approximately fifty years. In

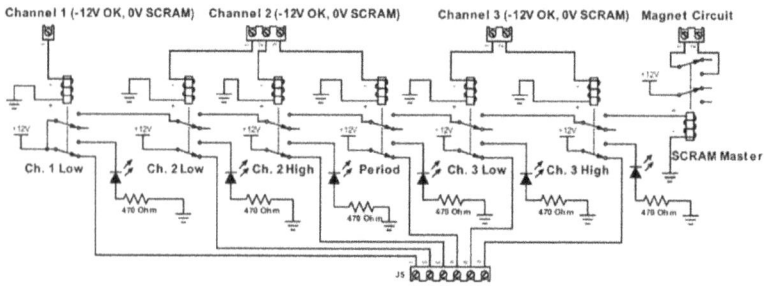

Figure 5.
SCRAM circuit arrangement [1].

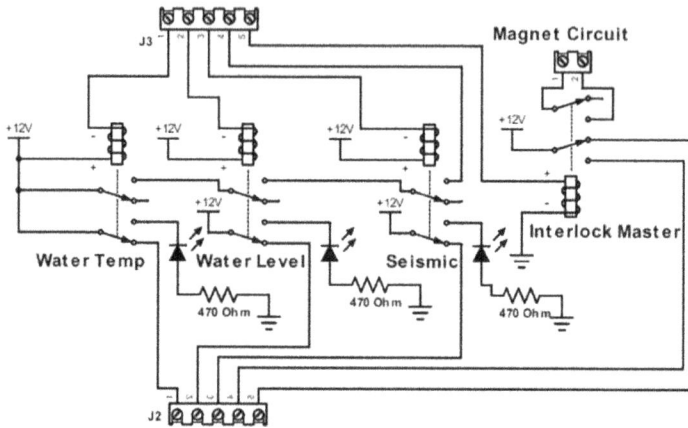

Figure 6.
Interlock circuit [1].

2020, the original console was replaced with an upgraded console (see **Figure** 7). The primary motivation for upgrading the console centered on the use of vacuum tubes for the SCRAM circuits in the old console. Obtaining replacement vacuum tubes became very difficult since these items are no longer manufactured in large quantities. The upgraded console uses solid state relays rather than vacuum tubes. In addition to the use of solid-state relays, the upgraded console has all new wiring, instrumentation, switches, and knobs.

While the AGN-201's core will essentially never be exhausted, support systems such as the instrument systems and their neutron detectors, reactor controls and control rod drives require periodic upgrading. The current financial state of universities and the perceived difficulty in conforming to regulatory requirements tends to encourage using the original 60-year-old tube-based control systems and other equipment until their failure rates leave no choice but to modernize. The cost of the engineering and manufacturing of upgraded instrumentation and equipment by outside firms can be too great for universities. Idaho State University recruited community volunteers with experience in project management and expertise in the design, construction, operation and startup of instrumentation and control, licensing of reactors and other relevant subjects for the university's second attempt to replace the original tube-based control system. The first attempt involved the design and construction of a complex, multiple-level printed circuit board that could not easily be modified. The second and successful attempt used a breadboard approach of circuit boards with holes that could be used to mount components. The

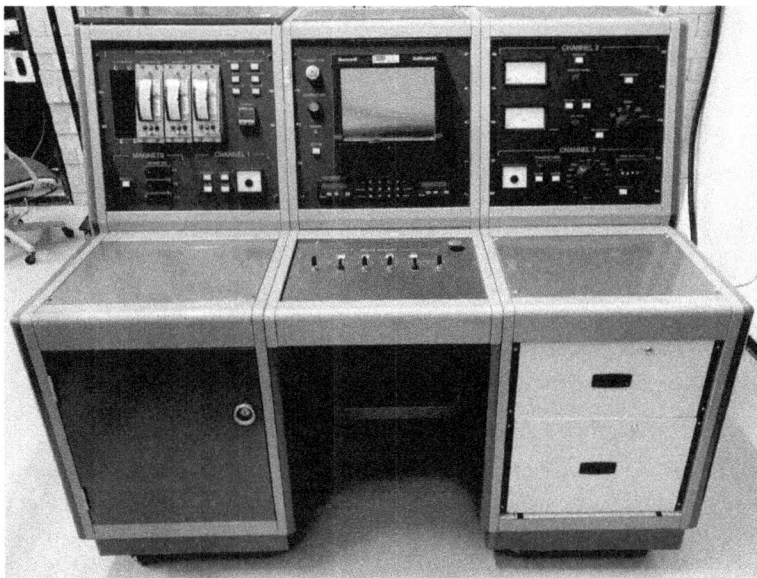

Figure 7.
Upgraded console installed in 2020 [1].

second attempt had very few changes to the design, a likely result of the lifetimes of experience of the community members in designing, repairing and maintaining analog systems. One of the main considerations was if the replacement system was to be analog or digital. The advantages and disadvantages of replacing the existing analog control system with a functionally equivalent analog system or attempting to replace it with a digital system were weighted. A replacement analog functional replacement appeared to be simpler and easier from a regulatory standpoint.

From a lifecycle cost standpoint, the analog system's lifetime was envisioned to be decades, whereas digital technology is rapidly advancing, and the lifetime of a digital system was envisioned to be a few years. Analog enjoys far superior cyber security than digital, and maintaining cyber security appeared to be an unnecessarily potential burden to the university. It was decided to replace the system under a 10 CFR 50.59 evaluation. The community expert in licensing helped write the 10 CFR 50.59 document and helped ensure applicable codes and standards had been followed. The community member also purchased and donated some of the components. Another community member and two graduate students worked with the community members to document the project in their thesis. One of the community members became the Project Manager and kept the project moving even during the height of the COVID-19 pandemic shutdown. He also reviewed the design and construction and assisted with troubleshooting. The collaboration of community and university personnel worked well to produce and complete the replacement instrumentation and had the time to transfer knowledge. It is anticipated that the same model will be applied to other modernization efforts going forward.

3. Capabilities

The nuclear instrument systems, convert the neutron flux at the detectors adjacent to the core into instrument readings that the operator interprets to control the core. Each part of the loop can be tested. The neutron flux at neutron detectors is often assumed by designers to be essentially constant at a given power level, whereas from very low to moderate levels of neutron flux, such as found during very low power and shutdown operation, the neutron flux can vary considerably in amplitude in a random manner. The random manner results from the characteristic random nature of the decay of neutron sources that supply the reactor with neutrons at low power and shutdown.

One consequence of the variation in neutron flux is that it appears as an unwanted variation in the display of a channel and might result in inadvertent period trips. Nuclear instrument channel 'noise' is generally considered an unwanted (and often misunderstood) variation in a signal. It can be electronic noise that is externally introduced to the circuit and must be minimized so it does not distort the true readings or the ability of the reactor operator to identify the average signal. It can also be due to the normal random decay of a neutron source, where it is a valid part of the signal. It can be very difficult to visually identify if the noise is due to the valid operation of the core, or if it is due to electrical interference.

A statistical test called the 'Chi-Squared' test can be applied to data from pulse-type channels such as the startup channel. A Chi-Squared test is often used at power reactors to verify the startup channels and any temporary startup-range neutron detectors used for loading fuel are displaying counts from neutrons rather than noise. The Chi-Squared test will identify if the noise is electronic interference or valid and due to a neutron source, although it will not identify the source of the electric noise.

The AGN-201 offers the opportunity for training and evaluating the nuclear instrument channels with a very low neutron signal, lower than typically encountered at commercial power reactors. The neutron flux at a detector must be low enough that the channels will display changes in signal (jumps) from individual neutron interactions, and the channels must support attaching a scaler-timer. The test is useful when the AGN-201 is shut down and a neutron source is supplying neutrons. If neutron flux is low enough, as it is when the neutron source is inserted, even the channel 2 and 3 ion chambers might be evaluated with the Chi-Squared test. In both cases, a scaler-timer is required to total the counts in a given time interval.

The AGN-201 offers a unique opportunity to explore the variations the current signal of a current neutron instrument channel without the time pressure and limitations on connecting test instruments at a power plant. A properly designed test can demonstrate that current signals from a neutron detector consist of a number of pulses of very small electrical charges, each resulting from the individual disassociation of B^{10} upon absorbing a neutron.

Teaching-reactors such as the AGN-201 provide the opportunity to measure a wide range of characteristics, and to gain experience and practice in conducting the same measurements that are performed at power reactors during low power physics testing following the loading of the first core, following refueling, and even during power operation to characterize the stability of the reactor. The tests generally involve changing a parameter such as reactor temperature or control rod position, and observing the corresponding change in rate of change in reactor neutron flux. Commercial reactors use a so-called 'Reactivity Computer' to infer the change in reactivity from a change in a parameter. The AGN-201 allows students to build, operate and evaluate the operation of analog and digital reactivity computers themselves [5, 6].

The AGN-201 could be used to evaluate and improve test procedures that would be used on future first-of-a-kind reactors, and to train future reactor engineers and other operating staff. In addition to gaining experience and practice in conducting the measurements, students can develop the skills required to write test procedures and to conduct high-quality test programs in a low-risk environment. The AGN-201 can also potentially offer realistic simulations of conditions at other reactors so newly written test procedures can be conducted and improved prior to being used at the reactor. Test engineers, reactor operators and others, including regulators, from other facilities can benefit from the training available at the AGN-201. The AGN-201 can be useful in observing the principles and some of the parameters being tested at other reactors, thereby allowing the test procedures to be validated and problems discovered.

The AGN-201 operates at very low power levels (microwatt range), often termed 'zero-power' where its operation closely resembles most other reactors when they are operated at low power levels, below the point of adding sensible nuclear heat. Even at power reactors, many of the core physics measurements that are made following refueling or core alterations occur with the core subcritical or with the reactor just critical on delayed neutrons at low power level. They include monitoring the core during shutdown operation, while core alterations during refueling are being made, during the approach to criticality, and reactor state point measurements and core physics parameter measurements in a suite of 'low power physics tests.' Some measurements are made at both low power and at-power, and only a few are restricted to high power operation. The AGN-201 is therefore capable of providing conditions for most of the core physics measurements found at power reactors [7].

The explanations and demonstrations of the theory and measurement techniques of subcritical core physics can be of interest to reactor physicists, instrumentation and control technicians and engineers, operators and managers of nuclear facilities, health physicists and criticality safety personnel. The phenomena of subcritical multiplication of source neutrons requires a 'multiplying medium,' neutron source and neutron detector. The common technique that is used at reactors is an 'inverse multiplication ratio' or '1/M' plot. The increase in count rate as control rods are moved in steps, and corresponding decrease in '1/M' plot are readily apparent. The plot is typically used to infer the point of criticality, in this case the position of the control rods. The reactivity of the AGN-201's control rods have been characterized well enough to illustrate the increases in count rate as positive reactivity is added. The demonstration can be relevant for power reactors to illustrate monitoring techniques during core alterations such as fuel loading and about establishing boron dilution warning setpoints. At pressurized water reactors with a soluble boron shim, the source range channels include the ability to establish a setpoint whose warning will help operators stop a dilution that could lead to an inadvertent reactivity change. The count rate at typical alarm setpoints can be low enough that the random variations in neutron production by the source becomes apparent. The resulting variation in source range channel readings, coupled with the requirement for a response time, can make it difficult to establish a setpoint that provides for enough warning but does not have false alarms.

Subcritical measurements to measure the values of parameters that formerly were measured during low power physics testing can save utilities considerable time and money. One is measuring the reactivity worth of control rods by raising and then dropping control rods, which can also be demonstrated in the AGN-201. Control rod drop times are also measured following refueling and other core alterations. The techniques and difficulties in measuring the positions of the controls during the drop, and the response of the nuclear instrumentation can be demonstrated in the AGN-201.

The state-point measurements of a reactor are measurements of parameters whose values define the operating condition, or 'state' of the reactor. Examples of parameters include reactor temperature and control rod positions are made to evaluate the reactivity of the reactor, and for comparison with core physics code predictions. Accurate state-point measurements are crucial in assessing the operation of the core and are made when the reactor is first brought critical after a refueling outage, and periodically throughout core life. The technique is simple and involves adjusting parameters such as control rod position, temperature and boron concentration in reactors with soluble neutron poison so the reactor is just critical at a given power level. A careful measurement, where reactor power is essentially constant, provides the best data. The AGN-201 allows operators and reactor engineers to explore their ability to establish just critical conditions, and to compare the measurements of parameters with calculations.

Low power physics measurements are conducted with a critical reactor whose power level is below the point of adding observable sensible nuclear heat, also known as 'reactors without feedback.' The measurements include the state-point measurement mentioned earlier, control rod reactivity worth, moderator temperature measurements, core stability measurements using a 'core oscillator' with variable, regular changes in reactivity, delayed neutron lifetime, irradiation of metallic foils to determine reactor power and more.

The operation of the AGN-201 is licensed and regulated by the Nuclear Regulatory Commission. The reactor and its conduct of operations are periodically inspected, particularly its documentation, and orderly documentation requires timely, accurate, truthful completion of forms, operating logs and more. Operating a nuclear reactor requires developing the valuable skills of discipline, focus and attention to detail, communication and more. The AGN-201 requires the same attitudes and abilities as higher-power test reactors. The full force of regulations is applied to the AGN-201. The opportunity to operate a nuclear reactor, regardless of size, is a unique experience that can benefit people who choose to put forth the time and effort. Students have opportunities to participate in a disciplined, regulated environment that is required of operators of a nuclear reactor that can shape their outlook on life and work ethic at a pivotal point in their lives. Students and other potential operators are invited to study, pass exams, and be responsible for the operation of a nuclear reactor providing a valuable and unique experience for those considering entering the field of nuclear power.

4. Conclusion

The Idaho State University AGN-201 reactor is a very safe, low-power, solid-core reactor designed with students and teaching in mind. It was developed in the late 1950's by AGN to satisfy the need of university nuclear engineering departments for a relatively inexpensive, safe, flexible and available reactor with a long design life. The AGN-201 reactor is well suited for teaching and research activities. The solid-core AGN-201 reactor requires no active cooling system, uses a simple shielding arrangement, and the very low operating power results in trivial burnup providing an operating lifetime exceeding many decades. The AGN-201 reactor is used to help prepare nuclear engineering and other students for entry into the nuclear workforce. The reactor introduces students to the disciplined, structured environment of operating a reactor licensed by the Nuclear Regulatory Commission. The reactor offers students hands-on experience, provides opportunities to demonstrate the operation of reactors and a variety of the traditional applications of reactors such as neutron activation, and introduces them to the application of nuclear instrumentation, applied principles of health physics, and more. With the recently installed reactor console upgrade, the ISU AGN-201 reactor is poised to serve students for many decades to come.

Author details

Chad L. Pope[1*] and William Phoenix[2]

1 Idaho State University, Pocatello, Idaho, USA

2 Idaho Falls Idaho, USA

*Address all correspondence to: popechad@isu.edu

IntechOpen

References

[1] Safety Analysis Report. Idaho state university AGN-201M research reactor. License No. R-110, Docket No. 50-284. 2003

[2] Skoda R, Peddicord K. Rebirth of AGN-201M: Practical ways of using the proven training reactor. In: 19th International Conference on Nuclear Engineering. Chiba Japan: ICONE19-44136; 2011

[3] Pope C. Prompt Neutron Decay Constant Measurement Using Rossi's α Method [Thesis]. Pocatello: Idaho State University; 1993

[4] Malicoat A, Pope C. Design improvements to the ISU AGN-201 reactor SCRAM, interlock, and magnet circuits. Annals of Nuclear Energy. 2020;**136**

[5] Levinskas D. Installation of an Automatic Reactivity Control System into the AGN-201 Reactor at Idaho State University [Thesis]. Pocatello: Idaho State University; 1990

[6] Smith C. Reactivity Measurement Software System for the AGN-201 Reactor Using Inverse Point Kinetics [Thesis]. Pocatello: Idaho State University; 1990

[7] Baker B. Comparison of Open Loop and Closed Loop Reactivity Measurement Techniques on the ISU-AGN-201 Reactor [Dissertation]. Pocatello: Idaho State University; 2013

Section 3

Select Reactor Analysis Techniques

Chapter 5

Core Reload Analysis Techniques in the Advanced Test Reactor

Samuel E. Bays and Joseph W. Nielsen

Abstract

Since becoming a national user facility in 2007, the type of irradiation campaigns the Advanced Test Reactor (ATR) supports has become much more diverse and complex. In prior years, test complexity was limited by the computational ability to analyze the tests' influence on the fuel. Large volume tests are irradiated in flux traps which are designed to receive excess neutrons from the surrounding fuel elements. Typically, fuel elements drive the test conditions, not vice versa. The computational tool, PDQ, was used for core physics analysis for decades. The PDQ code was adequate so long as the diffusion approximation between test and fuel element remained valid. This paradigm changed with the introduction of the Ki-Jang Research Reactor—Fuel Assembly Irradiation (KJRR-FAI) in 2015. The KJRR-FAI was a prototypic fuel element for the KJRR research reactor project in the Republic of Korea. The KJRR-FAI irradiation presented multiple modeling and simulation challenges for which PDQ was ill suited. To demonstrate that the KJRR-FAI could be irradiated and meet safety requirements, the modern neutron transport codes, HELIOS and MCNP, were extensively verified and validated to replace PDQ. The hybrid 3D/2D methodology devised with these codes made analysis of the ATR with KJRR-FAI possible. The KJRR-FAI was irradiated in 2015-2016.

Keywords: advanced test reactor, Ki-Jang Research Reactor, HELIOS, MCNP, 3D/2D methods

1. Introduction

In 2015, the advanced test reactor (ATR) began irradiations of the Ki-Jang Research Reactor—Fuel Assembly Irradiation (KJRR-FAI) test. Concurrent with the KJRR-FAI experiment program, the ATR was in the process of software quality assurance (SQA) for a more robust transport-based code, the Studsvik-Scandpower HELIOS code. The use of HELIOS enabled high quality (i.e., NQA-1) core reload and safety analysis of the ATR cycles for which irradiated the KJRR-FAI test.

The neutronic communication between the KJRR-FAI and the ATR fuel elements required 3D analysis. However, HELIOS is a 2D code. At the time, high fidelity 3D transport simulation of the ATR was too computationally expensive to be used for fuel reload and safety analysis. The solution of intra-plate power peaking in the ATR fuel elements was particularly challenging as this requires a significant number of particle histories in a Monte Carlo method and excessive mesh density in a deterministic transport method. As a workaround, the well-known Monte Carlo N^{th} Particle (MCNP) code was used to provide the axial peak-to-average power peaking factors which allowed for computationally efficient calculation of new core reload patterns that would satisfy the irradiation needs of the KJRR-FAI while ensuring safe operation of the ATR.

2. Background

2.1 The advanced test reactor

The ATR is a water-moderated, beryllium-reflected, pressurized water reactor with a serpentine arrangement of plate fuel taking on a four-leaf-clover likeness [1]. Each of the cloverleaves plus the center region are each referred to as lobes. The design power rating is 250 MW. However, it is currently operated at about 110 MW, and occasionally at powers approaching 250 MW to support higher power experiments. The metallic fuel plates consist of a highly enriched uranium (HEU 93wt% $^{235}U/U$) uranium-aluminide (U-Alx) dispersed in aluminum. This dispersion is sandwich clad in aluminum alloy, Al-6061. The fuel serpentine contains 40 fuel elements, each containing 19 curved plates. The fuel plates are swaged into side-plates forming the fuel element. The angular separation between the two side-plates is 45 degrees. The inner and outer four fuel plates contain natural boron carbide (B4C) to suppress radial (i.e., plate-to-plate) power peaking. The inner 11 fuel plates do not contain B4C. Also, the UAlx concentration is varied by plate to minimize radial power peaking.

Initial criticality as well as the power share among the lobes is maintained using hafnium plates on rotating control drums, called Outer Shim Control Cylinders (OSCCs). The burnup reactivity decrement is made up partly with OSCCs but also with annular hafnium neck-shims (i.e., 24 hafnium control rods) which are removed from the four aluminum neck arms in the center region. Numerous penetrations in the reflector and neck arms allow for non-instrumented "drop-in" as well as instrumented capsules, in addition to the nine flux traps. A picture of a typical reactor configuration is provided in **Figure 1**.

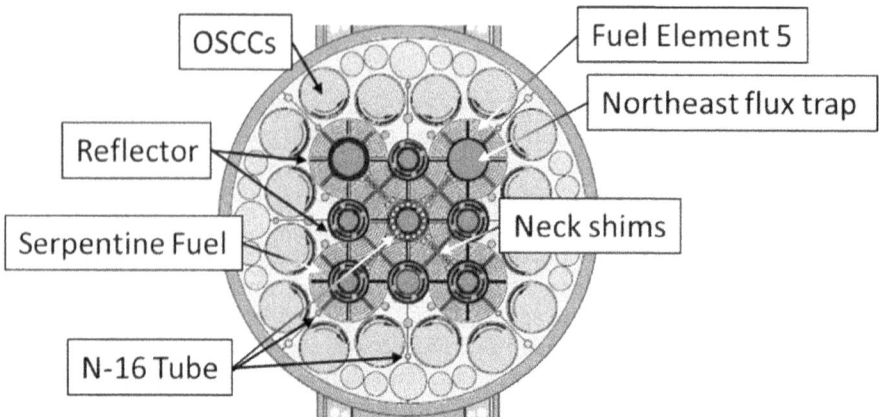

Figure 1.
The ATR basic configuration.

2.2 Ki-Jang Research Reactor—Fuel Assembly Irradiation

The Ki-Jang Research Reactor is a new isotope production reactor being pursued by the Republic of Korea [2]. This fuel is the first-of-a-kind of U-Mo fuel for commercial utilization; thus, it requires a license to be granted and a qualification of the fuel at scale. Thus, the KJRR-FAI is a full-size prototype designed to test mechanical integrity, geometric stability, acceptable dimensional changes, and assurance that the performances of the fuel meat and fuel element are stable and predictable during irradiation. This testing was conducted in the northeast lobe of the ATR from October 2015 to February 2017 in cycles 158A, 158B, 160A, and 160B [3]. These irradiations successfully demonstrated the KJRR fuel element's reliability in prototypic conditions to a burnup of 83.1% U-235.

The KJRR fuel element is based on the very successful plate-in-box fuel concept used in many research reactors across the world. Coincidently, this type of fuel has its origins in ATR's predecessor, the Materials Testing Reactor (MTR) [4]. The KJRR fuel is of the genre of high-density High Assay Low Enriched Uranium (HALEU) research reactor fuels, enriched to 19.75% ^{235}U/U [3]. The fuel meat is a dispersion fuel consisting of uranium-molybdenum alloy (U-7Mo) (i.e., seven w/o Mo) dispersed in an Al-5Si matrix (i.e., five w/o Si) (**Figure 2**). This dispersion fuel is clad in aluminum alloy Al-6061. There are 21 straight (not curved) fuel plates. The inner 19 fuel plates have a uranium density of 8.0 g-U/cm^3. The outer two fuel plates have a uranium density of 6.5 g-U/cm^3. The enrichment zoning is to reduce the radial power peaking in the fuel element (**Figure 3**).

The overall dimensions of the KJRR fuel element are 76.2 × 76.2 × 1010 mm. However, the active height of the KJRR-FAI fuel meat is only 60 cm (23.6 inch), which is less than half that of the active height of an ATR fuel element which is 48 in [5]. The KJRR-FAI was irradiated in the ATR northeast flux trap (**Figure 1**).

2.3 Computer codes

2.3.1 HELIOS

HELIOS version 2.1.2 is a general $x-y$ coordinate deterministic transport code. Arbitrary geometry is created by user defined nodes, connected to form line segments, then closed to form the spatial mesh. The code supports property overlays, such as composition, temperature, and density. These overlays are mapped to each mesh in the arbitrary 2D geometry. Geometry-corrected resonance integrals are calculated on-the-fly for every spatial mesh of the arbitrarily heterogeneous

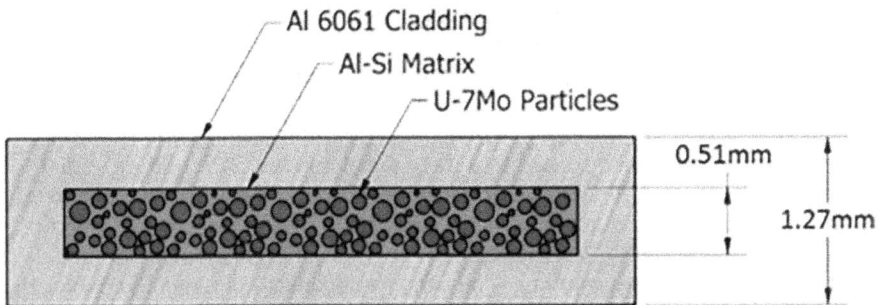

Figure 2.
KJRR-FAI fuel plate showing U-7Mo dispersion in Al-Si matrix.

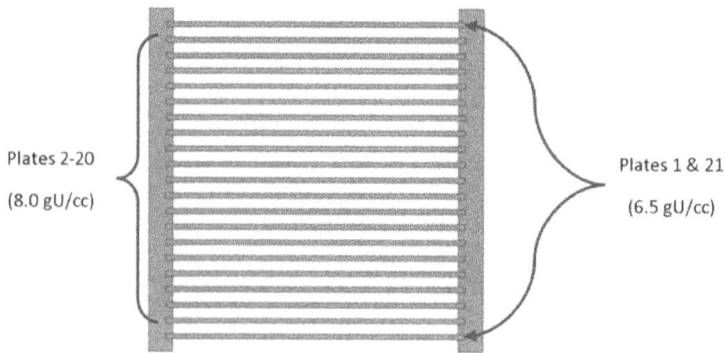

Plates 2-20

(8.0 gU/cc)

Plates 1 & 21

(6.5 gU/cc)

Figure 3.
Profile view of the KJRR-FAI prototype fuel element.

geometry description using the subgroup resonance treatment. HELIOS uses 49 groups derived from ENDF/B-VII. Very large and complex geometries are supported by subdividing the geometry into smaller subsystems. Each subsystem is solved explicitly via the collision probability (CP) transport solution, or method of characteristics (MOC), and then coupled with adjacent subsystems by sharing interface currents [6]. The angular dependence of the interface currents is discretized by a subdivision of outward/inward angles (i.e., a directional half-sphere). In the HELIOS model, all possible azimuthal directions crossing the interface are discretized into four equal sectors of equal weight. A HELIOS model containing the KJRR-FAI is provided in **Figure 4**.

From 2010–2015, the HELIOS code underwent extensive Verification and Validation (V&V) to elevate the software quality of HELIOS to Nuclear Quality—1 (NQA-1) [7, 8]. The HELIOS code replaced the neutron diffusion code, PDQ, for performing core reload analysis and associated safety calculations.

2.3.2 MCNP

MCNP uses the Monte Carlo method for solving particle (e.g., neutron and photon) transport in a continuous energy, angle, and three-dimensional space representation of the reactor core [9]. MCNP also makes use of ENDF/B-VII cross-section libraries. The Monte Carlo solution method represents particle interaction as probabilistic collisions between traveling particles and atomic nuclei. Therefore, the MCNP solution can be considered to be a near exact representation of reality to within the accuracy of the input nuclear interaction cross-section data. However, this level of solution fidelity comes at greater computational expense compared to a 2D deterministic code such as HELIOS.

MCNP 3D models, shown in **Figure 5**, of the ATR core with the KJRR-FAI loaded were developed for comparison with HELIOS. The MCNP model of an ATR fuel element consists of homogenized regions. The 19 fuel plates and associated coolant channels are homogenized into three radial regions. Each of these radial regions are partitioned into seven axial layers. Each axial layer is depleted separately during the depletion calculation. Though increases in computational speed is currently enabling 3D Monte Carlo solutions to be much cheaper, production calculations are still very time-consuming. Thus, the homogenization is done to reduce the required computational burden of resolving the geometry of every plate while still providing the desired level of accuracy for heating rates in the experiments. This is common practice when using MCNP to solve for heating rates in ATR experiments.

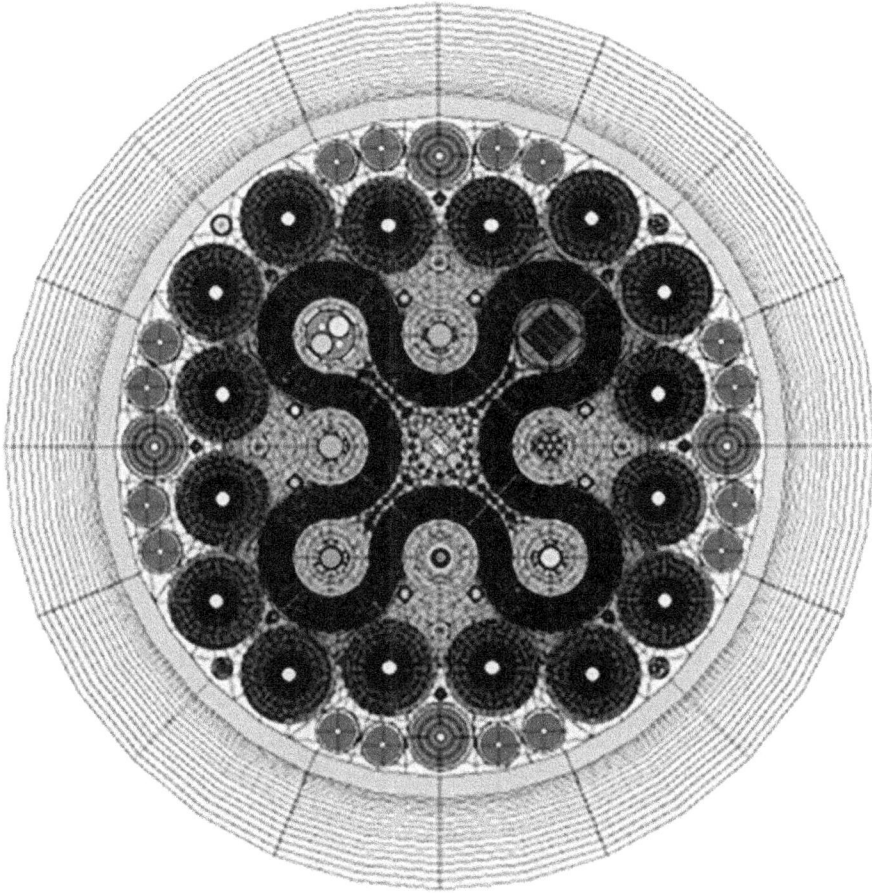

Figure 4.
HELIOS model of ATR (cycle 158A) with the KJRR-FAI loaded in the northeast flux trap.

Fuel depletion is solved using the ORIGEN2 code using the tallied neutron fluxes from each of the MCNP 21 regions. The ATR operating cycle is broken into discrete time-steps. MCNP solves for the one-group neutron flux and coalesced absorption and fission cross-sections in each of the 21 regions in each fuel element. These fluxes are passed to ORIGEN2. ORIGEN2 solves the Bateman equations to deplete the fuel. The depleted compositions are then passed back to MCNP.

MCNP is used to compute the axial peak-to-average peaking factor which is multiplied against HELIOS intra-plant powers during post-processing. The combination of the HELIOS 2D solution for every sub-plate region with the axial peak-to-average factor for every fuel element allows the final predictive core performance calculations to provide adequate 3D information. Typically, many design evolutions of different fuel loading patterns, and OSCC and neck-shim withdrawal patterns are needed to demonstrate the cycle's operating requirements can be met while respecting all safety limits. HELIOS is used for these design evolutions with axial peak-to-average factors provided by MCNP in the final design calculation.

The MCNP code was also validated against extensive fission wire activation measurements made in the advanced test reactor critical (ATRC) facility [5, 10]. The ATRC facility is zero power replica of the ATR used for low-power activation

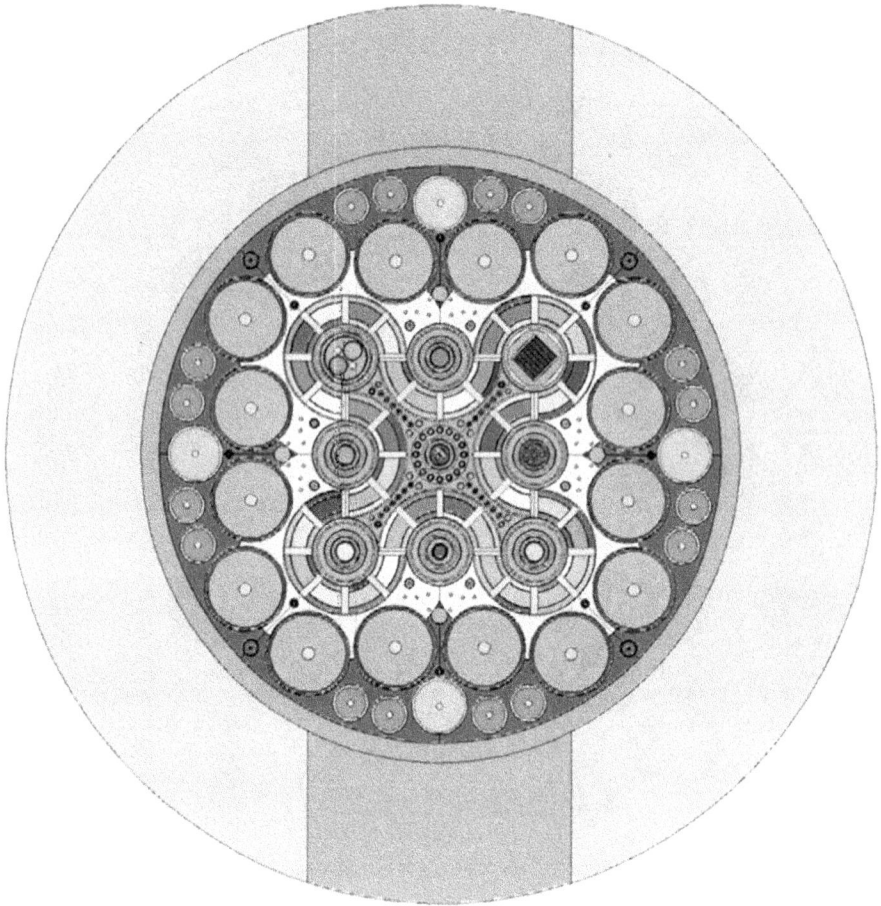

Figure 5.
MCNP model of ATR (cycle 158A) with the KJRR-FAI loaded in the northeast flux trap.

analysis to verify power distributions and to measure the reactivity worth of experiments prior to being inserted into the ATR.

3. Irradiating a square fuel element in a round flux trap

3.1 Managing core reactivity for high worth tests

Typically, ATR flux traps irradiate large volume experiments having high flux requirements. By definition, a flux trap is designed to 'trap' the excess flux from the surrounding fuel elements. This typically implies that the neutrons from the experiment do not greatly influence power production (or power distribution) in the ATR fuel elements. However, given the sheer quantity of fissile U-235 introduced by the KJRR-FAI, it became readily apparent that the KJRR-FAI could drive the northeast lobe power, rather than the northeast lobe driving the KJRR-FAI power. The thermal limits of the KJRR-FAI test could be exceeded unless the excess reactivity introduced by the KJRR-FAI could be managed. The KJRR-FAI test needed to be maintained at a power <2.3 MW to ensure its thermal margins could be met.

Several burnable poison options were investigated, but ultimately abandoned as this would interfere with the desired neutron flux environment.

It was decided that highly burned ATR fuel elements could be loaded into the northeast lobe to essentially "sponge" the reactivity introduced by the KJRR-FAI. However, even with the use of these highly burned fuel elements, the northeast quadrant of OSCCs would need to be rotated inwards to keep the northeast lobe at the desired 19 MW.[1] Typically, the burnup distribution of the fuel elements is selected such that the OSCCs can be rotated relatively (though not exactly) evenly. Said differently, it is desirable to manage power distribution around the serpentine using the fissile content of the fuel elements, not by using the OSCCS for power shaping. This does not always happen in practice, but this is a general goal of the fuel reload analysis. **Figure 6** shows the localized power distribution for 10 regions per plate for every plate in the core for cycle 158A. The OSCC are in the startup position at 29 degrees.

Rotating the northeast control bank to nearly "all-in" acceptably suppressed the northeast lobe's power but unacceptably robbed the whole core of excess reactivity. This required adding fresh assemblies somewhere else in the core such that the requested cycle-length could be met.

3.2 Updating power peaking factors

Prior to HELIOS, the 2D Cartesian mesh version of PDQ was routinely used to predict core reactivity, lobe-power distribution, and localized plate power peaking. However, the axial component of power peaking had been incorporated via an empirical correlation assuming the ATR thermal flux was a "chopped" cosine shape. The fresh fuel axial peak-to-average ratio was 1.43. The bounding thermal-

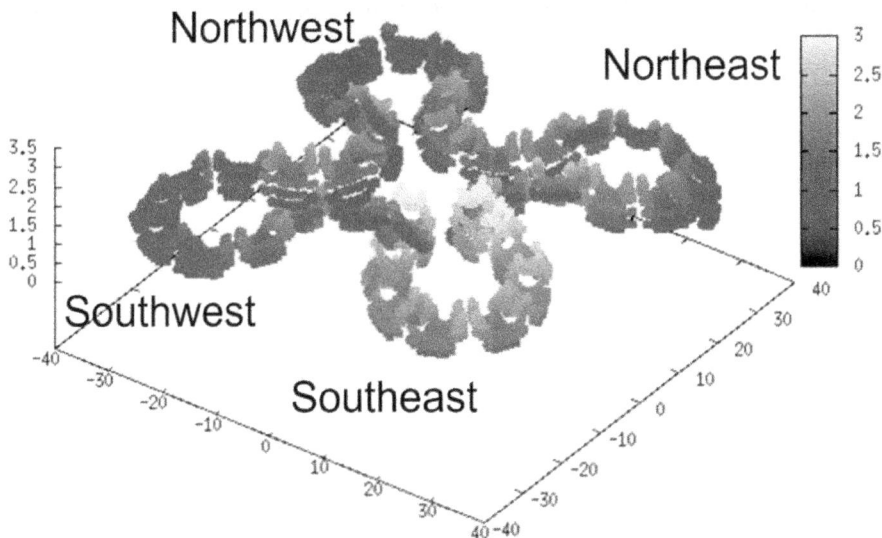

Figure 6.
The dimensionless point-to-average power density ratio for every fuel region in the HELIOS model for 'balanced' OSCCs at startup. Note, that the HELIOS 2D power density is corrected for axial power peaking using data from MCNP.

---

[1] The 19 MW for the eight ATR fuel elements in the northeast lobe, not including the 2.3 MW power from KJRR-FAI.

hydraulic analysis for the ATR assumes this axial power shape, originally calculated by PDQ, as universal for every cycle. Changing the axial power shape requires an update to the ATR thermal-hydraulic safety analysis, or at minimum a calculation to assure that the existing safety limits are not challenged by the new axial shape.

Prior to the KJRR-FAI, the chopped-cosine rule was rigorously enforced by ensuring that new experiments would not cause a major deviation from the established axial power shape.[2] An acceptance band is used to ensure that new tests would not violate the chopped cosine rule. Experiments that did not meet this criterion would require redesign if the chopped cosine rule could not be met. Typically, the MCNP code is used to design ATR experiments and used to predict the axial peak-to-average factor. If the MCNP analysis finds the axial peak-to-average power factor will likely be non-compliant to the chopped cosine rule, a measurement of the axial shape in the ATRC facility is considered to verify the calculated axial shape.

Note that the axial power peaking factor is significantly greater for the MCNP calculation with the KJRR-FAI in the northeast flux trap than it is for the chopped cosine rule. This is primarily contributed to the influence of the KJRR-FAI fuel loading being concentrated near the mid-plane, i.e., within ±30 cm about core-midplane. The axial peak-to-average power factor was calculated using MCNP by tallying the power in the fuel every two inches when the ATR fuel elements are all assumed to be fresh. This calculation was repeated with a generic test configuration typically used as an experiment backup in the northeast flux trap, called the Large Irradiation Housing Assembly (LIHA). The LIHA consists of an arrangement of cobalt and aluminum rods and is considered a standard backup for the northeast flux trap when not in use. The axial peak-to-average power factor in fresh ATR fuel elements was found to be ~1.5 when neighboring the KJRR-FAI and ~1.4 when neighboring the LIHA. The MCNP tallied power profile for fresh ATR fuel elements adjacent to the KJRR-FAI versus the LIHA is shown in **Figure 7**.

Modifying the test design was not an option for the KJRR-FAI; thus, the empirical chopped cosine shape would need to be rederived. Furthermore, the evolution of this power shape considering depletion effects would need to be considered. Even without the significant axial distortion due to the KJRR-FAI, fuel naturally depletes preferentially at mid-plane due to geometric shape or buckling of the neutron flux. This axial variation in burnup, and hence fuel nuclide distribution, needs to be represented

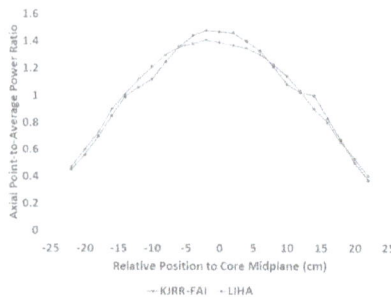

Figure 7.
Comparison of the axial power shape in ATR fresh fuel (Fuel Element 5, Coolant Channel 2) computed using MCNP due to the KJRR-FAI versus the standard LIHA northeast flux trap configuration.

---

[2] Control rods for gross reactivity control excluded from the ATR design as they would introduce an axial power tilt as a function of insertion depth. OSCCs could provide reactivity shim without significant change to localized power distribution, thus allowing constant flux conditions for the test locations [4].

in the HELIOS model (just as previously with the PDQ code). This axial burnup variation also impacts the axial power shape as represented in **Figure 8**.

Within the PDQ-based methodology, a three-dimensional extension of the "$x-y$" PDQ analysis was needed to compute the effect of the axial burnup shape on excess reactivity, axial power peaking, and axial burnup peaking. PDQ could be used to solve a 1D r-dimensional as well as a 2D $r-z$ coordinate system. These two features were used together to calculate 1D and/or 2D reactivity biases and axial multipliers due to fuel burnup. To derive a generic peak-to-axial power factor, a single lobe is approximated by a right circular cylinder (RCC) comprised of a generic in-pile tube (IPT) encircled by eight fresh ATR fuel elements. These ATR fuel elements are represented by seven fueled concentric annuli with no side-plates. The modelled seven annuli represented fuel plates 1, 2, 3-4, 5-15, 16-17, 18, and 19, respectively. Each fueled annulus is represented by a homogenized cell containing water, aluminum, and UAlx fuel matrix. The RCC lobe is also recast as a 1D r-dimensional model. Both the 1D r-dimensional and 2D $r-z$ model are depleted at 60 MW for 50 days, or 3000 MW$-$days (MWD) of "lobe-exposure". The 2D/1D reactivity bias and the axial power peaking factor are then set to a polynomial fit as a function of lobe-energy.

The PDQ axial peak-to-average factor is provided in Eq. (1).

$$A(t) = \frac{P_m(t)/V_m}{\sum P_i(t)/\sum V_i} = \frac{\mathcal{P}_m(t)}{\mathcal{P}_a(t)} \tag{1}$$

P_m represents power at midplane. This is the average power for regions of fuel on core-midplane. V_m is the volume of these regions. P_i and V_i represents the power and volume of all fuel mesh in the PDQ $r-z$ model. Note that the cursive, $\mathcal{P}$, represents power density. Time, t, represents fuel burnup in units of MWD, referred to as lobe-exposure. The fuel element axial burnup peaking factor is then derived from the indefinite integral of the axial power peaking factor.

$$B(t) = \frac{\int_0^t P_m(t)\,dt}{\int_0^t \mathcal{P}_a(t)\,dt} = \frac{\int_0^t A(t)\mathcal{P}_a(t)\,dt}{\int_0^t \mathcal{P}_a(t)\,dt} \tag{2}$$

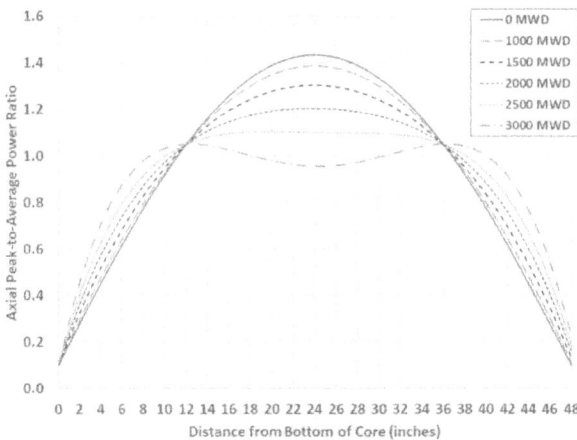

Figure 8.
Approximate axial power-to-average factors created as 2D/1D factors using the r–z PDQ RCC lobe model.

The average power density of the simple r-z model is held constant; thus, this factor may be eliminated.

$$B(t) = \frac{\int_0^t A(t)\,dt}{\int_0^t dt} = \frac{1}{t}\int_0^t A(t)\,dt \tag{3}$$

The behavior of $B(t)$ with depletion can be represented with a simple polynomial. In fact, for the duration of only one cycle, it is essentially linear.

$$A(t) = A_0 + A_1 t + A_1 t^2 \ldots A_n t^n \tag{4}$$

From basic calculus, the power rule can then be used to find the antiderivative of $A(t)$.

$$B(t) = \frac{1}{t}\left(A_0 t + \frac{A_1 t^2}{2} \ldots \frac{A_n t^{n+1}}{n+1}\right) \tag{5}$$

This process is simplistic but allows for accurate reproducibility of 3D power and burnup behaviors with burnup. This is true so long as power and burnup behavior in the x-y frame are separable from the axial-z frame. This is generally the case with ATR. This process was used to compute $A(t)$ and $B(t)$ for HELIOS using the MCNP code.

The beginning-of-cycle 3D/2D MCNP calculations occur just after the HELIOS fuel selection. With the core load pattern found, the MCNP 21-region model is created. The ORIGEN2 code is used to independently deplete each of the 21-regions assuming an approximate flux shape for an ATR fuel element. The depletion time is carried such that the sum of the 21 U-235 masses and the end of the depletion agrees with the U-235 inventory of that element used in the HELIOS model. The combination of 3D fuel nuclides, 3D experiment models, as well as OSCC and neck-shim positions as a function of burnup constitute the 3D MCNP model.

Unlike the 2D/1D peaking factor used in the PDQ methodology, the 3D/2D MCNP peaking factor may be derived for every fuel element. Here again, it is important to note that this method is very useful only when the $x-y$ frame is separable from the axial frame. **Figure 9** shows the change in the axial

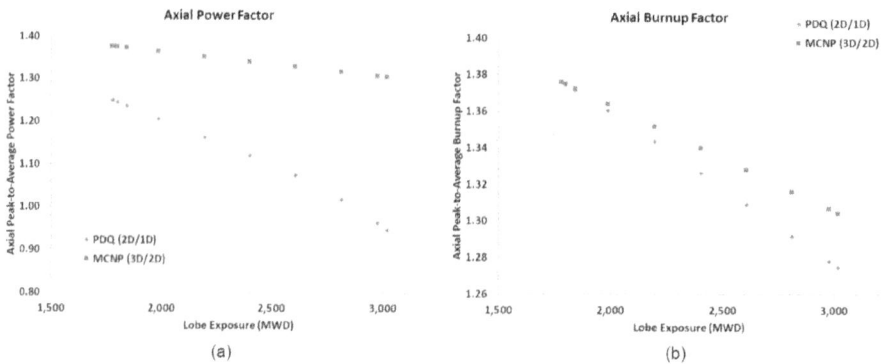

Figure 9.
Comparison of axial peak-to-average factors in previously irradiated ATR fuel elements: power factor (a) and burnup factor (b).

peak-to-average factor for fuel element five (shown in **Figure 1**) in the northeast lobe computed with MCNP compared to the generic PDQ factors.

The axial peak-to-average power factor is slower to change with burnup when the KJRR-FAI is present. This too can be attributed to the influence of the KJRR-FAI. It is noteworthy that because the KJRR-FAI is HALEU, as opposed to HEU, it has much more internal fertile-to-fissile conversion. This causes its own reactivity contribution to change slower with time. The KJRR-FAI generally drives the mid-plane power of the ATR fuel elements throughout the four cycles for which the test was irradiated. This caused an increase in the fuel burnup at mid-plane per assembly average burnup.

The combination of starting with heavily burned fuel elements to suppress lobe-power and the faster burnup rate of these elements required careful fuel element selection to ensure that the requested cycle-length could be achieved without exceeding the burnup limits on ATR fuel elements. Careful selection of fuel elements, essentially salvaging fuel elements slated for disposal, enabled the achievement of both the lobe-power and cycle-length constraints.

3.3 Finding a new equivalency

As mentioned previously, HELIOS assumes that all axial details are constant by nature of being a 2D code. This leaves the reactor analyst with one of two choices: extrude the most reactive axial region and assign this geometry and composition to the 2D HELIOS model (1), or axial homogenize all regions within the active core height and assign this composition to the 2D HELIOS model (2). For cases where it is important to preserve the overall reactivity worth of the experiment, its influence on lobe-power, and overall core reactivity, volume weighted axial homogenization is required. For cases where it is important to preserve the spatial self-shielding between the experiment and the nearby ATR fuel elements, extrusion is required. For KJRR-FAI, neither assumption could completely preserve the 3D behavior. Modeling the 21 fuel KJRR-FAI fuel plates as an extrusion artificially assigns the KJRR fuel density meant for 60 cm to the full 121.92 cm (48 in) active height of the ATR fuel. This would grossly over-estimate the fissile content of the test and artificially increase the reactivity contribution of the northeast lobe, thus producing a nonsensical estimate of required reactivity hold-down for the northeast OSCC quadrant. If the KJRR-FAI were homogenized with water and aluminum holders above and below it, the interplay between KJRR-FAI and ATR fuel element plates could be lost; thus, losing confidence in the burnup rate of the northeast lobe's fuel elements. The solution was a compromise between extrusion and homogenization.

By representing the KJRR-FAI mid-plane geometry in the HELIOS model, but reducing the uranium concentration in the fuel meat, the power of the test and its influence on power of the eight neighboring ATR fuel elements could be preserved. **Figure 10** shows the HELIOS computed KJRR-FAI power as a function of fractional uranium loading.

MCNP analysis showed that in order to keep the peak heat flux below the KJRR-FAI programmatic constraint (200 W/cm^2), the total fission power of the prototype fuel element would need to be kept to below 2.3 MW[3]. The minimum heat flux requested was 137 W/cm^2; thus, providing a lower bound of 1.6 MW. Therefore, the fractional uranium loading was reduced to 20% in order to provide a representative test power, as well as accurate power sharing behavior within the northeast lobe.

---

[3] The peak heat flux for the KJRR-FAI test program was 200 W/cm^2. The test itself in the ATR northeast flux trap had significantly more thermal margin.

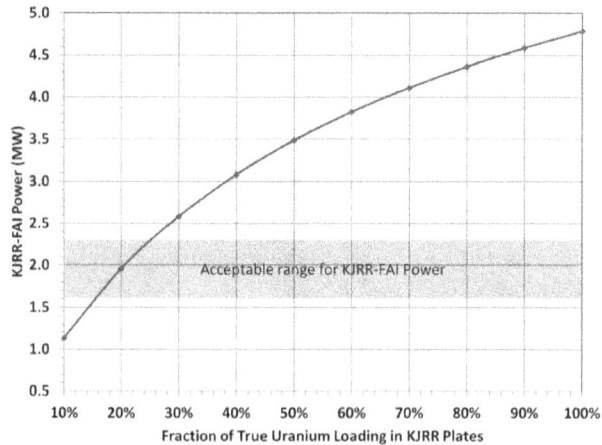

Figure 10.
As-modelled power, using HELIOS, of the KJRR-FAI.

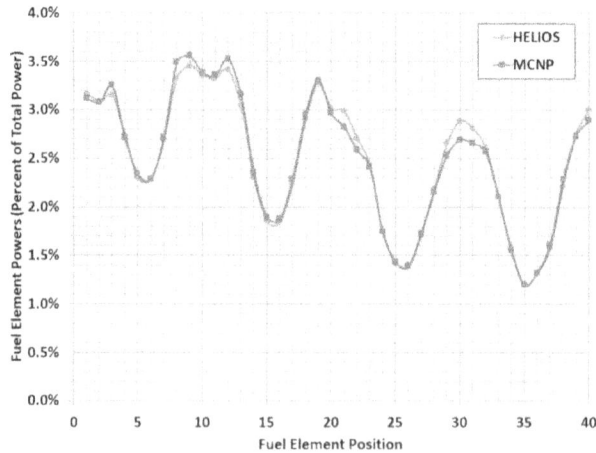

Figure 11.
Comparison of calculated fuel element power between MCNP with full 3D detail and HELIOS using a 2D model of the KJRR-FAI with 20% of the true uranium loading.

Note that the KJRR-FAI fuel meat height is roughly half that of the ATR fuel element, yet the fractional loading is only 20%. This is intuitive if one considers that fuel near the midplane has a greater importance (i.e., considering flux-weighting) compared to fuel further from mid-plane.

The adjustment process is verified against MCNP calculation of the 158A core with OSCC in the startup position of 29 degree rotated-out. This is referred to as the startup power distribution. The comparison of ATR calculated fuel element powers is shown in **Figure 11**.

This adjustment process is validated by the fact that the northeast lobe-power tracks well when compared to the lobe-power measurement system. ATR uses the water activation reaction, $^{16}O(^{1}n,p)^{16}N \rightarrow {}^{16}N$ ($T_{1/2} = 7.13$ s) $\rightarrow {}^{16}O + \beta^{-} + \gamma$ to indicate lobe-power. Ten flow tubes, two at central, four at ordinal positions near the lobes, and four at cardinal positions beyond the OSCC provide activation information to ion chambers. The ten ion chamber signals are converted via the least squares method to compute lobe-powers by a monitoring computer. This computer

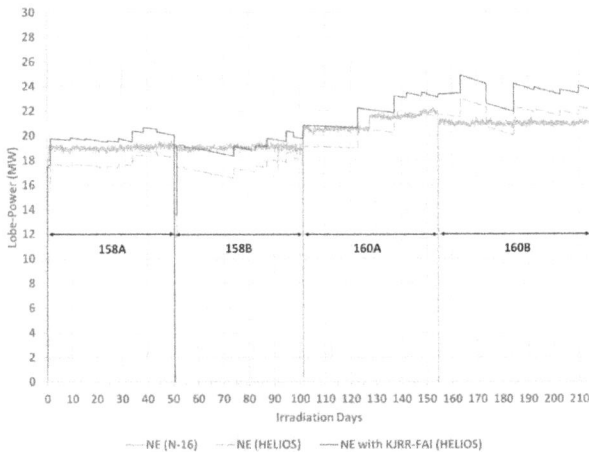

Figure 12.
Comparison of calculated versus measured (via the N-16 system) northeast lobe-power for ATR cycles: 158A, 158B, 160A, 160B.

also records the gross calorimetric power of the reactor, as well as the OSCC, and neck-shim positions every hour. This information is combined into a post-cycle analysis of the ATR cycle using HELIOS. This "As-Run" calculation serves two purposes. It provides accurate fuel element depletion results which are then tracked for fuel management records. It also serves as a continuing improvement process for code maintenance of HELIOS and associated ATR models. A comparison of calculated by HELIOS versus measured by the N-16 system for all cycles containing the KJRR-FAI is shown in **Figure 12**. In the figure, the calculated northeast lobe power is shown with and without the KJRR-FAI. The KJRR-FAI power can be calculated by HELIOS, as was shown in **Figure 10**. However, calculating experiment power is not part of the typical As-Run process. Therefore, unfortunately, this data is not available. However, the MCNP As-Run for which provides data to the KJRR-FAI project is available and is included in **Figure 12**.

4. Summary

The implementation of HELIOS as a design tool for core reload and safety analysis of the ATR is one of the first examples of using whole-core transport codes in such a capacity. Traditionally, codes such as HELIOS, are used for cross-section condensation for use with nodal diffusion codes. Since HELIOS is an arbitrary geometry code, this suits it well for creating cross-section datasets for much faster nodal diffusion codes that then analyze CANDU, RBMK, or VVER reactors. Generally, HELIOS would be an ideal code to support ATR fuel reloading analysis because the ATR was designed with minimum axial perturbation in mind; hence, the use of control drums over control rods. However, with the promise of higher solution fidelity has come more complex experiment designs. Since the KJRR-FAI cycles, more geometrically complex, high fissile worth, and/or high neutron absorber tests have been irradiated in ATR. The challenges of such experiments are as follows: finding fuel elements capable of providing the best irradiation conditions for all the customers of the ATR National Scientific User Facility (NSUF). In the case of high worth tests, this requires selecting fuel elements near the end of their life, but not so spent that they exceed their burnup limit by time the requested cycle-length has been reached.

Assuming this is possible, sufficient fuel must be loaded in the core to have sufficient excess reactivity to accomplish this cycle-length. However, this is limited by the amount of shut-down margin available in the OSCCs. If the loading is too rich, startup could occur in the non-linear range of the OSCC reactivity worth curve, risking a missed startup prediction. Not discussed here, but if a lobe is designed for much higher power operation, i.e., the core power is closer to the maximum rating of 250 MW, the margins to thermal-hydraulic safety limits can be challenged.

Finally, each time the chopped cosine assumption is challenged by such axial heterogeneity, the axial profile is calculated with MCNP, then measured in the ATRC, and ultimately triggers an update to the bounding thermal-hydraulic analysis using the new axial profile. Indeed, this was the case for the KJRR-FAI cycles. A new thermal-hydraulic limit for the ATR fuel element was derived such that the ATR's flow instability, departure from nucleate boiling ratio, and other safety limits under transient conditions would not be challenged by the KJRR-FAI's alternative axial profile.

The irradiation of the KJRR-FAI has essentially demonstrated that advanced codes can support advanced hardware. However, there is a tendency to believe that advanced codes can change the operating envelope of a nuclear reactor. The HELIOS and KJRR-FAI experience shows that the operating envelope is set by margins to the safety limits and that these margins are established by measurements. In the case of KJRR-FAI, these measurements were careful fission-wire measurements of the axial shape in the ATRC. The KJRRR-FAI test was a great success and required a great amount of teamwork among physicist and code developers that did the HELIOS code verification and validations, the reactor engineers who did the fuel reloading analysis, ATR plant operations who supported the fission wire measurements, and safety analysts who could understand the historical analysis with PDQ and connect those assumptions with modern application.

Author details

Samuel E. Bays* and Joseph W. Nielsen
Idaho National Laboratory, Idaho Falls, Idaho, United States of America

*Address all correspondence to: samuel.bays@inl.gov

IntechOpen

References

[1] Kim S, Schnitzler B. Advanced test reactor: Serpentine arrangement of highly enriched water-moderated uranium-aluminide fuel plates reflected by beryllium. In: International Handbook of Evaluated Criticality Safety Benchmarks, NEA/NSC/DOC (95)03/II, NEA-7231. Paris, France: OECD-NEA; HEU-THERM-022. 2014

[2] Seo C, Kim H, Park H, Chae H. Overview of KJRRR design features. In: Trans. of the Korean Nuclear Society Spring Meeting. Jangdae-dong, Korea: The Korean Nuclear Society; 30-31 May. 2013

[3] Kim J, Tahk Y, Oh J, Kim H, Kong E, Lee B, et al. On-going status of KJRR fuel (U-7Mo) qualification. In: European Research Reactor Conference, Brussels, Belgium: The European Nuclear Society, 14–18 May. 2017

[4] Stacy S. Science in the desert. In: Proving the Principle: A History of the Idaho Engineering and Environmental Laboratory, 1949-1999, Idaho Falls, Idaho, USA: Jason Associates Corporation, Idaho Operations Office of the Department of Energy, DOE/ID-10799. 2000. p. 3. Chapter 17

[5] Nielsen J, Nigg D. Physics validation measurements for highly-reactive experiment packages in the INL advanced test reactor. In: PHYSOR 2016, LaGrange Park, Illinois, USA: American Nuclear Society, 1-5 May. 2016

[6] HELIOS Methods (Version-2.0). Studsvik Scandpower, Idaho Falls, Idaho, 2009

[7] Bays S, Swain E, Crawford D, Nigg D. Validation of HELIOS for ATR core follow analysis. In: PHYSOR 2014, LaGrange Park, Illinois, USA: American Nuclear Society; 28 September-3 October. 2014

[8] Nigg D, Steuhm K. Advanced Test Reactor Core Modeling Update Project. Idaho Falls, Idaho, USA: Idaho National Laboratory, INL/EXT-14-33319; 2014

[9] Pelowitz DB, et al., MCNP6 Usure's Manual, Version 1.0. Los Alamos National Laboratory, LA-CP-13-00634, 2013

[10] Nigg D, Nielsen J, Norman D. Validation of High-Fidelity Reactor Physicss Models for Support of the KJRR Exerimental Campaign in the Advanced Test Reactor. Idaho Falls, Idaho, USA: Idaho National Laboratory, INL/EXT-42198; 2017

Chapter 6

Cyber-Informed Engineering for Nuclear Reactor Digital Instrumentation and Control

Shannon Eggers and Robert Anderson

Abstract

As nuclear reactors transition from analog to digital technology, the benefits of enhanced operational capabilities and improved efficiencies are potentially offset by cyber risks. Cyber-Informed Engineering (CIE) is an approach that can be used by engineers and staff to characterize and reduce new cyber risks in digital instrumentation and control systems. CIE provides guidance that can be applied throughout the entire systems engineering lifecycle, from conceptual design to decommissioning. In addition to outlining the use of CIE in nuclear reactor applications, this chapter provides a brief primer on nuclear reactor instrumentation and control and the associated cyber risks in existing light water reactors as well as the digital technology that will likely be used in future reactor designs and applications.

Keywords: cyber-informed engineering, nuclear digital instrumentation and control, digital instrumentation and control, cyber risk, nuclear cybersecurity

1. Introduction

Nuclear reactors rely on instrumentation and control (I&C) systems to maintain critical primary and secondary processes within desired parameters to ensure safe and efficient operation. Safety-related I&C systems are specifically designed to protect against critical failures that can lead to high consequence events. Designers rely on traditional safety-analyses, such as failure modes and effects analysis and probabilistic risk assessments (PRA), to inform them of specific protections needed in the design of these systems to maintain safe operation and the health and safety of the public.

I&C systems maintain real-time response, high availability, predictability, reliability, and distributed intelligence via a set of interconnected assets and subsystems that perform three main operations: acquisition, control, and supervision. Reactors have historically used analog I&C systems. As modernization occurs in the existing reactor fleet and as new advanced reactors are designed and commissioned, analog systems are replaced with digital I&C (DI&C) systems due to their many advantages, including reliability, efficiency, additional functionality, and data analytics. While DI&C provides enhanced operational capabilities, new risks associated with adverse impacts from cyber incidents are introduced. Whereas nuclear safety is the primary focus of reactor design, cyber risk must now also be considered in any digital-based reactor design. Cyber risk not only includes digital failures and unintentional cyber incidents, but the possibility that an adversary may try to purposefully disrupt, deter, deny, degrade, or compromise digital systems in such a manner as to place a reactor outside its intended design.

Since the complete set of failure modes for DI&C may never be fully known, and since DI&C can never be completely secured, a robust process is required to address and reduce cyber risk throughout the entire systems engineering lifecycle. Specifically, engineering and design personnel must be fully cognizant of the cyber risks and understand how to protect against intentional and unintentional cyber incidents. Cyber-Informed Engineering (CIE) is an approach in which cyber risks are considered at the earliest design stages and are continually reanalyzed throughout the entire lifecycle. Regardless of reactor design, cyber risk must be eliminated or reduced as much as possible to sustain a safe and secure nuclear industry.

The remainder of this chapter is organized as follows: Section 2 provides a background on nuclear reactor I&C systems, both analog and digital, as well as considerations for use of DI&C in new advanced reactor designs and applications. Section 3 steps through aspects of cyber risk analysis and cyber risk management for nuclear reactors. Section 4 provides an overview of CIE along with detailed descriptions of each CIE principle prior to concluding the chapter in Section 5.

2. Background

Chemical, manufacturing, and nuclear processes rely on instrumentation, such as pressure, temperature, and flow sensors, to measure and monitor process parameters. These industrial processes are then maintained by control systems that operate physical equipment, such as valves, pumps, and heaters, to keep the process parameters within predefined limits. Nuclear reactors vary by type (e.g., pressurized water reactor, pool-type reactor, liquid metal cooled reactor, molten salt reactor, gas cooled reactor) and purpose (e.g., power reactor, research reactor, nuclear propulsion). The remainder of this section first describes the fundamentals of generic nuclear reactor I&C prior to discussing the transition to digital technology, including its benefits and challenges. The section concludes with an overview of future DI&C applications, including those in new and advanced reactors, as well as integrated systems and decision support systems.

2.1 Fundamentals of reactor instrumentation and control

Nuclear reactors initiate and control nuclear fission or fusion reactions. These processes must be monitored and closely controlled to ensure reliable and efficient operation while maintaining the health and safety of the public. The number and type of parameters monitored in a reactor will vary depending on the reactor type and purpose, but both nuclear and non-nuclear instrumentation will likely be used. Nuclear instrumentation includes detectors to monitor neutron and gamma flux for routine reactor monitoring and control as well as reactor safety. Neutron detectors, such as proportional counters and ion chambers, are commonly used to provide source range, intermediate range, and power range monitoring, while gamma detectors are used for post-accident monitoring. These detectors may be out-of-core or in-core, or a combination thereof, depending on the reactor type. Other compact in-core detectors, such as small fission chambers or self-powered neutron detectors, are also commonly used for continuous real-time monitoring of reactor core conditions, including reactor power distributions.

Non-nuclear instrumentation includes sensors used to monitor process parameters, such as temperature, pressure, differential pressure, level, and flow.

Additionally, non-nuclear instrumentation may be used to monitor other parameters, including control rod position, area radiation, fuel-pin fission gas pressure, vibrations, acoustics, fuel or vessel strain, process fluid chemistry, moisture and gas analysis, and leaks.

Local instrumentation data is transmitted from the sensors to control board indicators, data recorders, applications, and control systems via analog or digital circuits, often through multiplexers or combinatorial logic circuits. Applications are commonly used to auctioneer (e.g., signal selection), aggregate, and/or perform calculations on the data to provide real-time reactor and plant status indications to operators.

While operators will also perform manual actions on a reactor, such as starting and stopping pumps or opening and closing valves, I&C systems are commonly used to automatically control reactor operations and maintain reactor safety. Control systems can be simple, like a single programmable logic controller, or complex, like a reactor control system. Control systems can combine numerous sensors, transmitters, controllers, and actuators to change the physical state of process equipment, such as a valves, pumps, or motors, by using signal feedback loops to monitor and maintain desired conditions. In a nuclear power plant (NPP), non-safety control systems may include feedwater control (or fluid control), turbine control, and reactor control.

Most nuclear reactors will have at least two types of control systems—reactor control systems and reactor safety systems. Depending on a reactor's purpose, there may also be other control systems, such as plant control systems in an NPP or experiment/sample control systems in a research and test reactor. Reactor control systems are used to control the nuclear fission or fusion reaction within specified acceptable fuel design limits by adjusting physical components according to the reactor design. For example, a reactor control system may raise or lower control rods in a light water reactor (LWR), turn control drums in a heat pipe reactor, or start or stop feedwater flow in a research reactor.

In an LWR, reactor control systems are used to maintain desired thermal megawatts by balancing primary and secondary systems. For example, as shown in **Figure 1**, an integrated control system may automatically maneuver reactor, feedwater, and turbine systems to match megawatts generated to megawatts demanded by adjusting control rod positions, valve positions, and pump speeds.

In comparison to reactor control systems, reactor safety systems are used to shut down and maintain safe shutdown of a reactor in the event a reactor safety limit

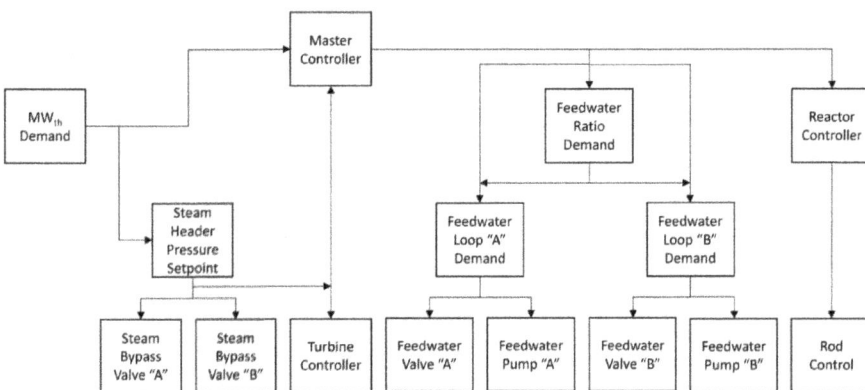

Figure 1.
A notional automatic integrated reactor control system for an LWR.

is met or exceeded to assure reasonable protection against uncontrolled release of radioactivity. For example, reactor safety systems in an LWR include Reactor Protection Systems (RPS), Engineered Safety Feature Actuation Systems (ESFAS), and diverse actuation or diverse trip systems. Reactor safety systems often use either two-out-of-four or two-out-of-three logic. For instance, an RPS may have four redundant instrumentation channels that monitor key parameters, such as reactor power, reactor coolant temperature, reactor coolant pressure, reactor coolant flow, reactor building pressure, reactor pump status, and steam generator level. If any design limits are exceeded on two separate channels, an automatic trip signal is sent to the control rod system to shut down the reactor. A notional representation of an RPS is shown in **Figure 2**.

In an LWR, an ESFAS is designed to provide emergency core cooling for the reactor and to reduce the potential for offsite release of radiation. Comparable to an RPS, ESFAS uses multiple channels of equipment in two-out-of-three logic (or similar) to monitor signals such as reactor coolant pressure and containment pressure. Based upon the specific coincident actuation signals received, ESFAS will start the required safety system, such as emergency core cooling systems, emergency feedwater, containment isolation and ventilation, containment spray, or emergency diesel generators.

2.2 Digital instrumentation and control

Although the first closed-loop industrial computer control system was installed by Texaco Company at its Port Arthur refinery in 1959 [1], I&C in nuclear reactors

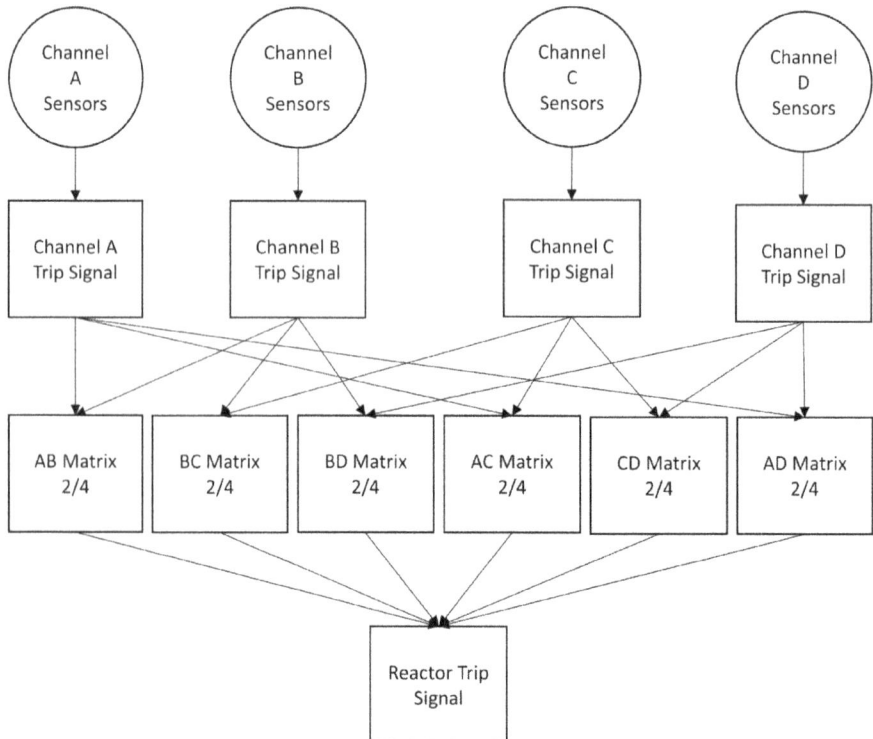

Figure 2.
Signals for a notional RPS.

largely remained analog until about 30 years ago when digital transmitters, indicators, controllers and data recorders began replacing analog sensors, indicators, actuators, and pen-based chart recorders. And, while non-safety digital control systems (e.g., feedwater control systems, turbine control systems and reactor control systems) are now commonly installed in nuclear reactors, safety-related digital control systems (e.g., RPS, ESFAS) are much less common, especially in the United States. The United States has been slow to adopt digital technology because of previously unanalyzed risks associated with new and unknown attributes, including common cause failures and cyber risks. International adoption of digital technology in nuclear reactors, including safety-related control systems, has been more aggressive than in the United States. Of course, new advanced reactors are being primarily designed with DI&C.

As described by the Nuclear Energy Institute (NEI), an I&C device in the U.S. power reactor industry is typically considered 'digital' if it contains any combination of hardware, firmware, and/or software that can execute internally stored programs and algorithms without operator action [2]. Hardware includes microelectronics, such as digital or mixed signal integrated circuits, as well as larger assemblies, such as microprocessors, memory chips, and logic chips. Hardware may also include other peripherals, such as expansion drives or communication controllers. Software includes operating systems, platforms, and applications used for process control, human machine interfaces, and other specific programs used for device or system operation. Firmware is software stored in non-volatile memory devices that provides low-level control specific to the hardware. Firmware executes higher-level operations and controls basic functionality of the device, including communication, program execution, and device initialization.

Field sensors and controllers may be standalone, small local systems, or larger distributed control systems. Devices may be connected by physical cables or wireless technology (e.g., WiFi, cellular, satellite, Bluetooth, radio frequency identification). There is also a range of communication protocols used in DI&C depending on the design and manufacturer.

The systems, structure, and components (SSCs) used in U.S. NPP safety-related protection systems are categorized as Institute of Electrical and Electronics Engineers (IEEE) class 1E technologies as defined by IEEE 308-1971 (and later) [3]. They must be designed to conform with General Design Criteria (GDC) in 10 CFR 50 Appendix A [4], IEEE 279-1971 [5], IEEE 308-1971 [3], and IEEE Std 603-1991 [6], as applicable based on construction permit dates. Guidance in Regulatory Guide 1.152 [7] and IEEE 7-4.3.2-2003 [8] may also be used to comply with Nuclear Regulatory Commission (NRC) regulations. Internationally, applications or components that perform IEC category A safety-related functions may fall under IEC 61513 [9], International Atomic Energy Agency (IAEA) SSR-2/1 [10], and IAEA SSG-39 [11] requirements.

These general design criteria include conformance requirements for independence and single-failure criterion such as defense-in-depth, diversity (i.e., different technology), redundancy (i.e., secondary equipment that duplicates the essential function), physical separation, and electrical isolation. The purpose of single-failure criterion is to ensure no single failure of a component interferes with the safety function and proper operation of the safety system [6]. Generally, it is impossible to prove that digital systems are error free. And, while common-cause failures can occur with analog equipment, it is more likely that software errors will result in common-cause failures, such as identical software-based logic errors that could cause simultaneous functional failure of all four RPS divisions. Thus, since unanticipated common-cause failures are more likely in digital systems than analog systems, there is increased burden to prove to the regulator that the design adequately meets the general design criteria outlined in the applicable requirements.

2.3 Benefits and challenges of digital instrumentation and control

The systems engineering lifecycle for analog modifications, such as changing mechanical relay logic, can take significant time to design, procure, reconfigure, and test hard-wired devices installed inside control cabinets. These changes can require many hours for maintenance personnel to rewire, physically rearrange components, and/or add new cabinets, terminal blocks, power supplies, and wiring. Labor resources are also required for post installation quality checks.

Contrary to analog I&C, a significant benefit of DI&C is the ability to quickly reprogram the functionality of a device or system with minimal physical hardware changes. These modifications are performed via microprocessors, expansive memory storage, and standardized communications that allow for remote connectivity. Moreover, the utilization of reusable software and common microprocessors lowers overall product costs. Moreover, the global supply chain has promoted further innovation, improved efficiencies, better product availability, and reduced costs.

An additional benefit of DI&C is the capability to incorporate numerous functions within one device. This capability reduces overall size of the I&C systems (e.g., fewer racks and cabinets) and relieves potential space constraints within facilities. Furthermore, the ability to choose from a wide array of functions in one device not only reduces the cost, but also allows for unique control algorithms not necessarily available in the past. Whereas analog I&C was limited to using a single proprietary signal conveying only one piece of information (e.g., the process value), adding a digital signal overtop an analog signal allowed for increased device diagnostics and calibration capabilities without any additional hardware changes and helped pave the way for logical extension of DI&C in nuclear facilities.

Other applications enabled by DI&C include enhanced online monitoring for condition-based maintenance systems. These systems improve visibility into equipment conditions to improve maintenance activities and potentially reduce or eliminate required preventive maintenance. Additionally, training departments are now able to simulate plant operations with fine detail that was difficult to achieve before.

On the other hand, digital technology introduces new challenges. As existing nuclear reactors are modernized, plant personnel throughout the organization must be trained on their design, installation, operation, and maintenance. This skillset is often very different than what is required for analog I&C and can take many years to acquire. Moreover, not only is there an increase in common-cause failures and potentially unknown failure modes with DI&C, but there is also additional risk associated with malicious and unintentional cyber threats not typically seen with analog I&C. These DI&C cyber risks are further described in Section 3.

2.4 Future technology considerations

2.4.1 New and advanced reactor designs

While existing reactors primarily designed and built with analog technology are transitioning to DI&C, new generation III+, small modular reactor (SMR), microreactor, and advanced reactor designs will likely apply digital technology from project inception to take advantage of increased flexibility, better performance, and improved reliability. It is anticipated that these designs will also include hybrid approaches, similar to existing reactors, incorporating both analog and DI&C components and systems for reactor control and reactor safety. However, since most of the new reactor designs will likely incorporate passive safety features, they may have fewer (or no) safety-related control systems compared to current LWRs.

Nuclear reactors are primarily designed with safety as the underlying principle. Ensuring safety of reactor personnel and maintaining the health and safety of the public is more important than secondary objectives, such as producing electricity or medical isotopes. Thus, any new reactor technology that challenges the nuclear safety paradigm is met with strong caution. However, as new advanced reactors are designed with DI&C, significant effort and analysis will be undertaken to ensure cyber risks are fully understood such that the designs will fully withstand regulatory and public scrutiny and not interfere with reactor safety. Nevertheless, the inclusion of passive safety features that reduce the footprint of digital safety systems not only reduces the number of high-consequence design basis accidents (DBAs), it also reduces overall cyber risk.

Sites built with multiple reactor modules (e.g., SMRs) may have additional I&C systems to enable integrated and coordinated operation across multiple units. Furthermore, proposed advantages of SMRs and microreactors include the capability for remote and autonomous (or nearly autonomous) operation, including anticipatory control strategies to maintain operational limits for both planned and unplanned internal or external disturbances which increase overall operational flexibility. The passive safety systems in advanced reactors may enable fewer operators and more automation, however, these new modes of operation and previously unanalyzed consequences require careful evaluation by designers and regulators to ensure minimization of cyber risks. Mobile reactor designs must also anticipate and address additional requirements for safe and secure transportation.

Similar concerns exist for remote operations, which is under consideration for advanced reactors in isolated environments or reactors connected to microgrids using autonomous distributed energy control schemes. Remote operations imply some finite distance between reactor and operator utilizing digital communications for both monitoring and control. Not only does the external pathway potentially enable an exploitable pathway for adversaries, it also potentially presents unanticipated cyber risks from communication failures.

2.4.2 Integrated energy systems

Whereas remote and autonomous reactor operation may have a long timescale for development, regulatory acceptance, and construction, integrated energy systems may be available on a shorter timescale. As shown in **Figure 3**, integrated energy systems use the thermal heat from reactors for other purposes, such as hydrogen generation, district heating, water purification, and chemical manufacturing. They may also have direct electrical connections to integrated systems. The interconnections between a reactor and these secondary processes will likely contain additional sensors, controllers, and actuators in order to balance the electrical and heat demands of the plant with the demands from the integrated energy systems.

2.4.3 New supporting applications

Digital twins are virtual replications of a physical system that can be used to provide various capabilities and decision-support at a nuclear facility. The degree of representation by a digital twin depends upon the computing power and the ability to accurately model both reactor physics and data-driven processes. Proposed applications include the use of digital twins for running artificial intelligence or machine learning (AI/ML) applications for hybrid control schemes, such as flexible operation for electric grid load-following, anticipatory control, or autonomous control; the use of AI/ML on digital twins for equipment condition monitoring, diagnostics,

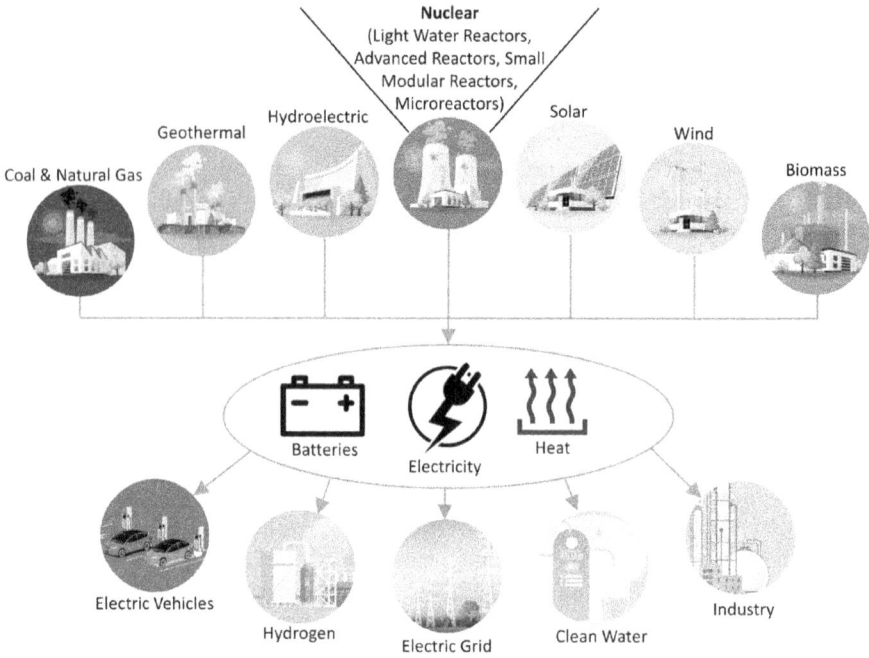

Figure 3.
Conceptual integrated energy system including generation sources and applications.

prediction, and prognostics; and the use of digital twins for designing engineering modification prior to building the actual physical system.

Using digital twins for reactor and/or system design may enable vulnerability discovery, such as potential for equipment failure, process anomalies, human error, or cyber compromise. Understanding system operation as well as potential vulnerabilities and consequences prior to construction is not only a benefit to designing better and safer reactors but also, if used with CIE principles as described in Section 4, a reactor with reduced cyber risk.

Applications of digital twins will likely continue to expand. The capabilities of digital twins, AI/ML, and other monitoring and control systems will be enabled with the increased use of wireless technologies (e.g., Wi-Fi, radio frequency identification, Bluetooth, Zigbee, cellular) in addition to traditional wired networks. Moreover, the use Internet of Things (IoT) or Industrial Internet of Things (IIoT) will continue to expand within nuclear facilities enabling improved efficiencies, reduced maintenance, and real-time insights for decision-making. Whereas the difference between operational technology (OT) and information communications technology (ICT) is that OT uses digital devices to control physical processes, such as nuclear reactors, IIoT uses a wide range of lower cost sensors that are traditionally connected via wireless networks to increase the number datapoints available for machine-to-machine communication and enhanced monitoring using data analytics, big data, and AI/ML.

3. Cyber risk

Risk is classically defined by Kaplan and Garrick as the possibility of loss or injury, including the degree of probability of such a loss [12]. Traditional safety PRA in the nuclear industry uses a logical framework combining fault tree analysis and event tree analysis to identify the likelihood and consequence of severe accidents which could lead to radiation release impacting the health and safety of the public. Nuclear safety PRAs typically use data on functional failures (i.e., manufacturer failure analyses, historical plant and industry failure data) along with known events (i.e., historical data on prior nuclear-significant events).

Unfortunately, the PRA approach is insufficient for cyber risk analysis as the complete set of failure modes for digital assets and systems may be unknown as they can fail in unexpected ways. Additionally, deliberate actions, such as intentional, intelligent, and adaptive actions by an adversary are challenging, if not impossible, to effectively model. Furthermore, threats and vulnerabilities are constantly evolving, a reality which does not lend itself to PRA. Therefore, rather than follow the Kaplan and Garrick risk triplet of 'scenario, likelihood, consequence' [12], cyber risk is better identified by evaluating threats, vulnerabilities, and consequences [13].

It is important to note that cyber risk includes all risk from both intentional and unintentional actions. Holistic cyber risk includes human performance errors and equipment failures as well as adversarial events. Adversarial events include malicious actions, including those by an unwitting insider, intended to cause damage or disruption to reactor and facility operations. Adding to the concern, intelligent threat actors can potentially adversely impact nuclear DI&C by remotely exploiting vulnerabilities, a threat that does not exist with analog I&C.

3.1 Consequence analysis

A nuclear reactor has a licensing basis that identifies high-consequence DBAs that can potentially lead to radiological release. This licensing basis includes those safety-related SSCs that must remain functional during a DBA to protect the health and safety of the public. While safety-related impacts are the primary concern, consequences from a cyber incident at a nuclear reactor could potentially range from intangible impacts (e.g., reputation damage, industry perception) to financial impacts (e.g., lost generation, equipment damage, repair costs) to adverse public health and safety impacts due to radiological release or theft of special nuclear material (SNM). Examples of low to high consequence impacts from a cyber incident are illustrated in **Figure 4**. **Table 1** expands on several of these consequences to provide causal examples of functional failures from hypothetical cyber incidents.

Cyber-induced consequences at a nuclear reactor can be minimized by maintaining availability, integrity, and confidentiality of DI&C components and systems. Nuclear reactors may be designed to run continuously (e.g., NPP) or intermittently (e.g., research and test reactor). In either case, data and communication flow must remain available to ensure safe and reliable operation of the reactor. Delay, disruption, or prevention of data or communication within an OT system can result in

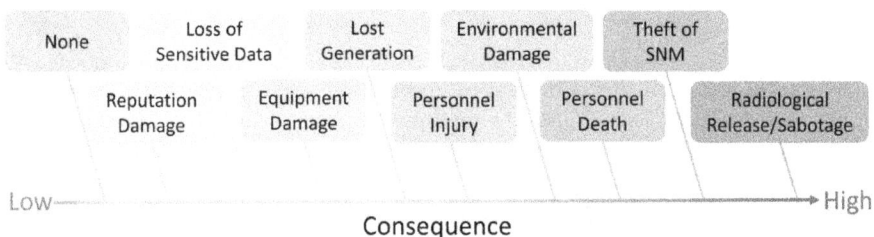

Figure 4.
Potential consequences from a cyber incident at a reactor.

Potential consequence	Functional failure	Initiating cyber incident
Radiological release	Failure of a safety system to actuate when needed	Digital RPS does not trip the reactor on low feedwater flow
Lost generation	Inadvertent actuation of a safety system leads to extended plant shutdown	Digital high-pressure coolant injection actuation with no loss of coolant
Lost generation	Inappropriate operator action	Operator does not recognize that a digital indicator on the main control board is incorrect
Equipment damage	Pump suction valve closed	Digital valve controller closed valve with pump running

Table 1.
Potential consequences from a cyber incident at a reactor along with hypothesized functional failure and initiating cyber incident.

unintended control actions, such as inadvertent component actuation or reactor trip. As listed in **Table 2**, cyber incidents that impact availability can be malicious and intentional, such as from a denial of service attack [14], or non-malicious and unintentional, such as excessive network traffic from failing equipment [15].

The integrity of DI&C information, data, and system parameters must also be maintained. Control systems require accurate, truthful, and complete information for safe and reliable operation. For instance, unintended modification of data, logic, or commands by man-in-the-middle attacks can cause equipment failure [16] or poorly executed software updates can reset plant data and cause actuation of a safety system [17]. Operators also rely on truthful and accurate data for decision making; inaccurate data on indicators or human-machine interfaces could cause operators to make improper decisions or perform incorrect actions. Operationally, it is often more dangerous to have a reactor in an unknown state instead of safely shut down. Consider an unexpected cyber incident that is visible to the operator—the operator can detect and respond to the incident, thereby minimizing further impacts. On the other hand, cyber incidents that are invisible to the operator can potentially result in persistent and higher consequence adverse impacts as operators are unaware of true reactor status.

While not as important in OT systems, confidentiality is also a cybersecurity objective. Loss of confidentiality, such as unauthorized exfiltration of sensitive information [18] or inadvertent posting of sensitive data in the public domain, can enable development of further attacks or cause other business-related concerns. Gaining sensitive nuclear information can provide adversaries roadmaps, schedules, vendors, plant layouts, and a host of other sensitive information shortening the attack timeline and delivering potential pathways to be considered towards ransomware, blackmail, or general political unrest.

3.2 Threat analysis

Cyber threat vectors into a nuclear reactor include wired and wireless networks or connections, portable media and maintenance devices (e.g., USB drives, maintenance laptops), insiders, and the supply chain. Furthermore, cyber threats can be classified as non-malicious or malicious. Non-malicious actions are often caused by employees or other facility personnel who perform actions not intending to cause harm. These actions are often human performance errors in which a worker mistakenly performs an adverse action, such as misconfiguring a device, selecting the wrong option, or disclosing sensitive information.

Security objective	Malicious incident	Non-malicious incident
Availability	Denial of service attack [14]	Failing equipment leading to excessive network traffic [15]
Integrity	Man-in-the middle attack [16]	Software update resetting plant data [17]
Confidentiality	Reconnaissance attack leads to data exfiltration [18]	Sensitive data posted on external site

Table 2.
Examples of malicious and non-malicious cyber incidents by security objective.

Malicious threats against nuclear reactors are initiated by adversaries with the intent to cause harm. Adversaries include recreational hackers, malicious and unwitting insiders, criminals, terrorist organizations, and nation states. Sophisticated attacks against nuclear reactors will likely be launched by organizations that have greater resources (e.g., skilled personnel, funding, time) and sufficient motivation (e.g., economic gain, military advantage, societal instability). Additionally, cyber-attacks may be one-dimensional or multi-dimensional, hybrid, coordinated attacks combining multiple threat vectors in both physical and cyber domains. For instance, adversaries may use cyber means to gain access to enable physical destruction or theft of SNM or use physical means to gain access to computer systems to enable unauthorized theft of sensitive information or sabotage.

In the United States, power reactors licensed by the NRC must provide high assurance that critical digital assets (CDAs) are protected against cyber-attacks, up to and including the design basis threat (DBT) [19]. CDAs are defined as digital assets associated with safety-related, important-to-safety, security, or emergency preparedness functions as well as support systems and equipment which, if compromised, would adversely impact these functions. A DBT describes adversarial attributes and characteristics, including level of training, weapons, and tactics, that must be defended against to safeguard the reactor against radiological sabotage and prevent theft or diversion of SNM. Generally, a beyond-DBT, a threat from an adversary who has capabilities beyond what is defined by the DBT, is considered nation-state activity which falls under responsibility of the state (e.g., federal government) for prevention, detection, and response.

3.3 Vulnerability analysis

Vulnerabilities are known or unknown weaknesses. Vulnerabilities in hardware, firmware, and/or software can leave digital assets susceptible to accidental failure or unintentional human error. Additionally, vulnerabilities may be exploitable, enabling adversaries to extract information or insert compromises allowing unauthorized access to perform malicious activities. Vulnerabilities can allow adversaries to penetrate and move throughout systems without the user's knowledge to compromise the availability, integrity, and confidentiality of complex control systems.

Most digital devices can be reprogrammed or modified to perform unintended or undesired functions. Any vulnerability that allows an unauthorized reprogramming or modification of a critical digital asset can result in adverse function of the DI&C systems. As most design approaches wait until system implementation to evaluate vulnerabilities, vulnerability response and mitigation often relies on bolted on security controls. However, if engineers who design and maintain complex control systems are trained to identify, understand, and mitigate these vulnerabilities throughout the lifecycle, including during design stages, vulnerabilities can be addressed early and often, thereby leading to lower overall cyber risk.

From a maintenance perspective, manufacturers often identify vulnerabilities and send information notices to asset owners along with mitigation measures, if applicable. Numerous vulnerability tracking databases and notification services also exist which serve to improve awareness and facilitate mitigation or protection [20–23]. Engineers and stakeholders should maintain awareness of these external vulnerability notifications or sites for their digital assets throughout the entire lifecycle so that they can be addressed immediately.

3.4 Cyber risk management

Of course, cyber risk cannot be calculated by simply multiplying numerically derived values of threats, vulnerabilities, and consequences together. For instance, low-threat, high-consequence cyber incidents will likely have a much different risk significance at a nuclear reactor than a high-threat, low-consequence incident. While many techniques have been proposed for incorporating the results of consequence, threat, and vulnerability analyses into a final cyber risk analysis [13], determining, evaluating, and prioritizing cyber risk is highly dependent on the reactor design, regulatory requirements, and organization's risk tolerance.

Cyber risk management is the continual process of analyzing cyber risk, evaluating and prioritizing the identified risk against organizational and regulatory requirements, and then applying risk treatments. In the United States, current nuclear power reactors typically follow guidance in NRC Regulatory Guide 5.71 [24] or the NEI cybersecurity series (NEI 10-04 [2], NEI 08-09 [25], and NEI 13-10 [26]) to identify CDAs and risk treatments. Corresponding cyber security guidelines for the international nuclear community are provided in IAEA Nuclear Security Series (NSS) No. 13 [27], NSS 17-T (Rev. 1) [28], NSS 42-G [29], NSS 33-T [30], and IEC 62645 [31]. For risk management activities, IAEA NSS 17-T (Rev. 1) refers readers to ISO/IEC 27005. Additionally, IEC 62443-3-2 provides an international security risk assessment standard for I&C systems [32]. Cybersecurity regulation and guidance for advanced reactors is still in development.

Regardless of the equation or formula used, cyber risk is managed by analyzing the potential worst-case consequences and then using risk treatments (e.g., avoidance or elimination, mitigation, transference, or acceptance) to lower the risk to a level acceptable to the organization. Unlike analog I&C, where failure analysis was the primary focus of PRA, the use of DI&C has resulted in the capability for hardware, firmware, and software to be altered in a manner not intended by the original design. Since both malicious and unintentional actions can potentially adversely impact operational functions, continually evaluating cyber threats, vulnerabilities, and consequences in a cyber risk management program is necessary to maintain awareness into the constantly evolving risk environment. The goal of this consequence-driven analysis is to prioritize risk treatments for those DI&C components needed to ensure critical reactor functions are maintained.

Consequence-driven, Cyber-Informed Engineering (CCE) is a formal cyber risk management approach that focuses on reducing the impact from high consequence events (HCE) for an overall business entity [33]. As shown in **Figure 5**, CCE is a four-step process. In phase 1, HCEs are identified and prioritized using a severity score calculated based upon consequence criteria weights and criteria severity. For the identified HCE(s), a system of systems analysis identifies the most critical functions in phase 2 and potential cyber-attack scenarios on those functions are then identified in phase 3. In phase 4, appropriate protection and mitigation strategies are developed.

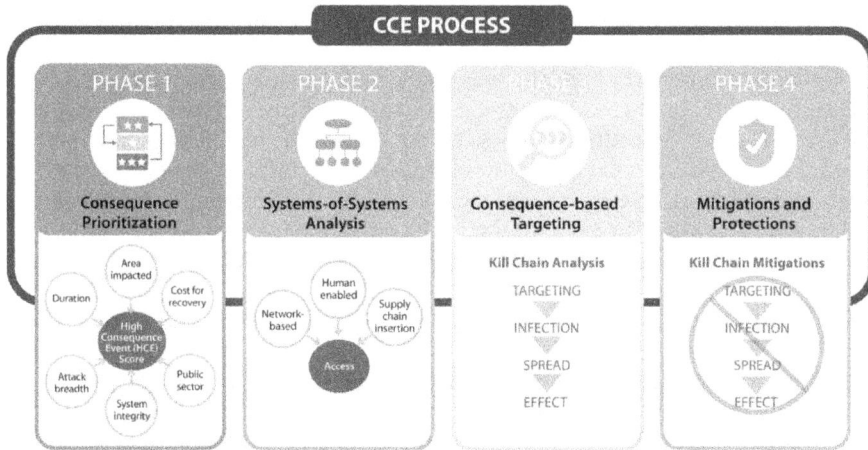

Figure 5.
The four-phase CCE process [34].

Additionally, cyber risk management must not only be considered for nuclear reactor SSCs, but also any digital technology used in their design, operation, and maintenance. For instance, AI/ML and digital twin applications are susceptible to both adversarial and unintentional cyber risk. These technologies are often considered 'black box' techniques in which the end-user is unaware of how the insights are determined. Even if more 'gray box' techniques are used, trust in AI/ML and digital twin models must be established to gain acceptance and approval by operators and regulators. Similarly, adversaries can gain access to these tools and cause data and/or model corruption to adversely affect model operation.

4. Cyber-informed engineering

Digital technology will be increasingly used in both existing and future nuclear reactors. While DI&C enables improved operations and new capabilities, the cyber risks must not only be understood, but risk treatments and protections must be put in place to lower this risk from malicious and unintentional actions. Whereas significant strides have occurred with securing ICT systems, these ICT-based solutions are not always effective for OT systems which are often designed to perform a limited set of functions and therefore have limited processing, memory, storage, retrieval, and proprietary communication protocols. Additionally, cyber risk mitigations have historically been applied after DI&C systems are installed, which limits the range of risk treatments available. On the other hand, applying the concepts of CIE throughout the entire systems engineering lifecycle can reduce overall cyber risk.

Engineers, operators, maintenance personnel, and other technical staff who support the systems engineering process are critical to the design, implementation, and secure operation of complex control systems. Nevertheless, this staff often lacks the necessary knowledge, skills, and abilities to effectively address and mitigate cyber risk. Given the critical functions of DI&C in nuclear reactors, this gap must be filled. For this reason, the Department of Energy Office of Cybersecurity, Energy Security, and Emergency Response is developing a national strategy for CIE to fundamentally change the culture of the engineering discipline to consider cybersecurity as a fundamental design principle.

4.1 Systems Engineering Lifecycle

Figure 6 illustrates the typical stages in the systems engineering lifecycle. While this model is intended to be used iteratively and potentially out-of-order throughout the lifecycle as design modifications occur, the left side of the V-model indicates a top-down approach moving from system to subsystem to component levels and the right side indicates a bottom-up approach through implementation, integration, and testing. This model is useful for both new builds (e.g., new reactor designs) or existing builds (e.g., engineering modifications).

4.2 CIE overview

CIE is a multidisciplinary approach that advocates the use of CIE principles in each of the systems engineering lifecycle stages to ensure that cyber considerations are included in every aspect of design, testing, implementation, operation, main-tenance, and disposal or decommissioning [36]. CIE is fundamentally a cyber risk management tool that complements existing OT cybersecurity risk standards and guidelines by incorporating engineering solutions along with ICT and OT cyber solutions to minimize risks from malicious and unintentional cyber incidents. Considering cyber risk and cyber risk treatments early and often throughout the lifecycle provides simpler, more secure solutions at lower cost, precluding the need to use ineffective, bolt-on solutions during later lifecycle stages.

As shown in **Figure 7**, the primary CIE principle that encompasses the entire CIE methodology is cyber risk analysis. The remaining CIE principles are divided into two categories: design principles and organizational principles. The CIE design principles are fundamental engineering design practices and techniques that build cybersecurity and cyber-resilience into DI&C early in the systems engineering lifecycle and then continue to ensure cyber-awareness is maintained throughout the remaining stages. This secure-by-design approach is more effective and less expensive than bolting on security controls after installation as the design can be influenced by factors that improve the ease, simplicity, and effectiveness of cyber considerations without impacting the performance of the intended system function.

Cyber risk is also reduced by instilling cyber-awareness at organizational- or facility-level functions. CIE organizational principles are fundamental cyber prac-tices that enable holistic integration of cyber considerations into other programs within the facility, such as asset inventory, supply chain, response planning, and training.

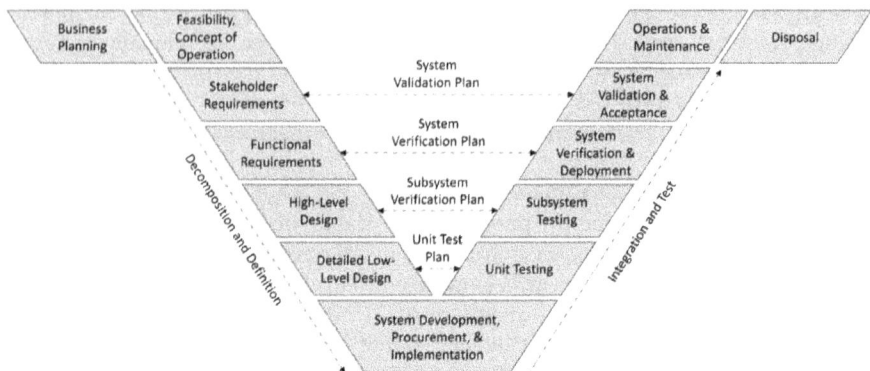

Figure 6.
Systems engineering V-model [35].

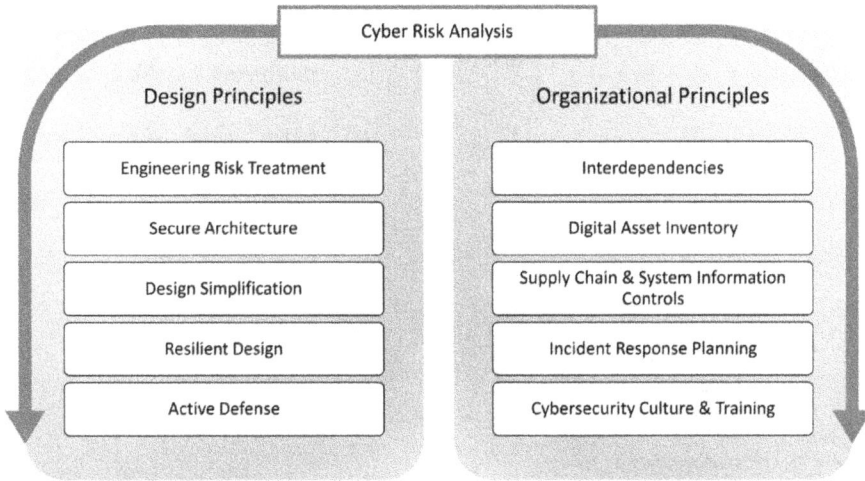

Figure 7.
CIE principles adapted from [36].

4.3 Cyber risk analysis

While cyber risk and cyber risk analysis were discussed in Section 3, it is important to remember that consequence-driven risk analysis is necessary to prioritize design requirements and risk treatments of those digital SSCs required for ensuring reactor safety and the health and safety of the public. Like the CCE methodology, since resources are often limited, organizations should first ensure that the most stringent protections are around those critical functions that, if compromised or lost, could lead to unacceptable radiological consequences, sabotage, or theft of SNM.

4.4 CIE design principles

4.4.1 Engineering risk treatment

Risk management is the process of identifying, evaluating, and responding to risk. Traditional risk treatments for responding to risk include risk avoidance or elimination, risk transference, risk mitigation, and risk acceptance. As shown in **Figure 8**, engineering risk treatments for cyber risk are similar, where risk can be designed out, shifted to another organization, mitigated with security controls or countermeasures, or accepted by making a conscious decision to tolerate the risk without implementing changes.

Security controls, as identified by National Institute of Standards and Technology (NIST) SP 800-82 [37], NRC Regulatory Guide (RG) 5.71 [24], or NEI 08-09 [25] are typically considered administrative, physical, or technical. As indicated in **Figure 8**, these controls mitigate cyber risk that cannot be eliminated. Unfortunately, engineering risk treatments, including security controls, are typically not considered until after installation. However, waiting until after installation is often too late to provide adequate protection. On the other hand, implementing engineering risk treatments during design stages can actually eliminate specifically identified risks by designing it out altogether or more efficiently and effectively reduce risk by incorporating security controls into the design.

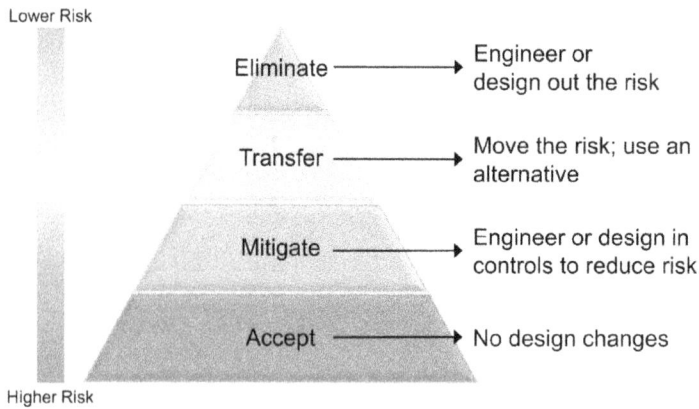

Figure 8.
Engineering cyber risk treatments [35].

4.4.2 Secure architecture

The goal of the secure architecture CIE principle is to establish network and system architectures that segregate and limit data flows to trusted devices and connections within and between subsystems, systems, and systems of systems. Properly designed architectures reduce cyber risk by isolating critical functions, minimizing the cyber-attack surface, and lowering the probability of unauthorized access or compromise of critical SSCs.

To ensure defense in depth, the design should consider use of isolated (e.g., air-gapped) or segregated network levels and zones, boundary devices, data flow rules, and unidirectional, deterministic communication, such as data diodes. In the United States, NRC Regulatory Guide 5.71 recommends power reactors to implement a defensive architecture with only one-way data flow from safety and security network levels outward to the plant network [24]. Internationally, as illustrated in **Figure 9**, the IAEA recommends implementing security levels with common requirements and zones separated by decoupling devices, such as data diodes and other boundary devices, such as gateways, routers, or firewalls, to minimize communications to untrusted devices [38]. Engineers should consider these secure architecture approaches during design stages to limit overall risks from compromised pathways or devices.

4.4.3 Design simplification

A cyber incident can only adversely impact DI&C functions if a vulnerability is exploited by a threat (intentional or unintentional). Vulnerabilities decrease as the complexity of DI&C decreases. Thus, the goal of the design simplification principle is to reduce the complexity of the system, component, and architecture while maintaining the intended function. Design simplification minimizes vulnerabilities and reduces overall cyber risk.

Design simplification is considered in conjunction with the secure architecture, resilient design, and engineering risk treatment principles. Complex or overbuilt designs result in a digital footprint larger than necessary. As the number of digital assets increases in a system, the number of digital failure possibilities and exploit locations also increases. Additionally, it is possible for adversaries to repurpose unused or latent functions and features on SSCs to behave in unanticipated ways.

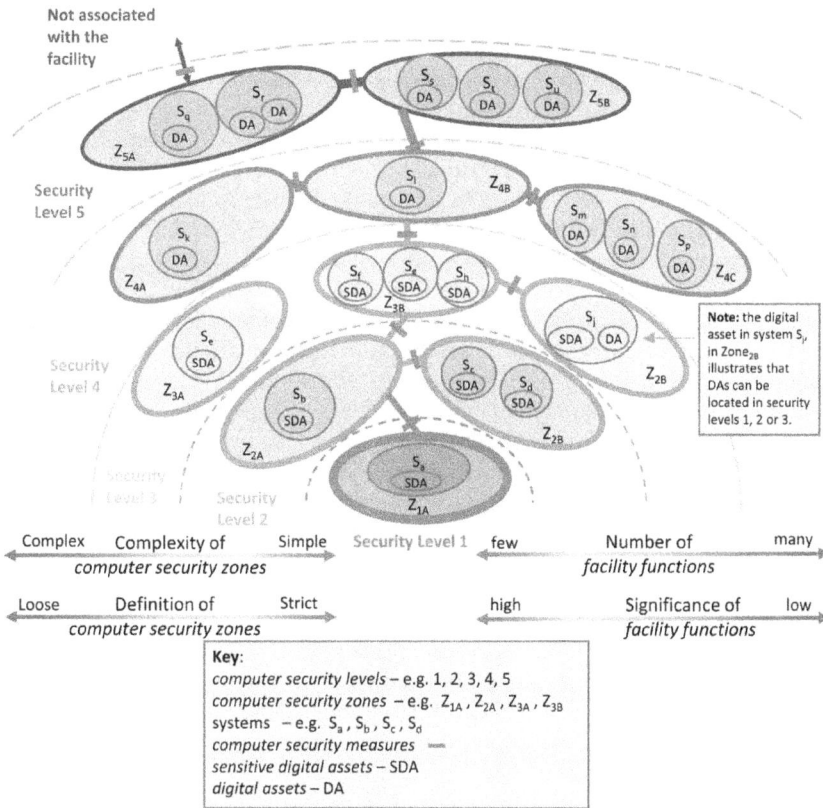

Figure 9.
Example implementation of a secure architecture [38].

On the other hand, simplifying the design, such as by using simpler digital devices or hardening the system by eliminating, limiting, or disabling unnecessary functions or capabilities, minimizes the overall cyber-attack surface and reduces vulnerabilities. The intent of design simplification is to simplify the engineering design itself, not sacrifice security requirements for the sake of simplicity. Nevertheless, in cases where extreme safeguards are required, analog I&C may be implemented instead of DI&C to protect against cyber incidents.

4.4.4 Resilient design

Resilience is a system's capability to withstand internal and external disruptions, including equipment failure, grid disturbances, or cyber incidents. A control system is resilient if it continues to carry out its mission by providing its required functionality despite disturbances that may cause disruptions or degradation. In nuclear reactors, general design criteria of separation, redundancy, diversity, and defense in depth are used for designing safety-related systems. Separation and independence are achieved by physical separation and electrical isolation. Redundancy is achieved by using more than one component to perform the same function. Diversity is achieved by using different technology within the system and with the redundant components.

Current DI&C systems operate in an untrusted environment, which presumes that users, devices, and systems cannot be trusted (e.g., users can be unauthorized, devices can be infected with malware). Additionally, it is impossible to design DI&C systems to withstand every malicious or unintentional cyber incident. Thus, resilient design is required to ensure continued safe and secure operation of the reactor and facility not only during an incident, but afterward as well.

While safety-related DI&C systems in nuclear reactors should be designed using the general design criteria, consideration should be given to designing similar features into non-safety DI&C systems to address this zero-trust paradigm, depending on the cyber risk prioritization. The objective of resilient DI&C design is to ensure continued operation of critical functions when possible, or graceful degradation when not possible, in the event of an SSC failure or cyber incident. Failure of one function, device, or system should not result in failure of another function. System design and control logic should attempt to eliminate the possibility of such cascading failures.

Additionally, resilient design may also include contingency planning and situational awareness. Contingency planning provides alternative methods for continued operation of critical functions. Using techniques, such as network and system monitoring, to provide situational awareness enables rapid decision making that may be needed for continued operation during a cyber incident. Moreover, operators have been trained to trust their instruments and indicators. This training model may need to be revisited due to the new zero-trust environment.

Finally, it should be noted that while resilient design may seem contrary to design simplification, the intent is to ensure that critical functions remain operational during a cyber incident. If additional devices are required to adequately assure resilience, there may be a tradeoff between resiliency and simplicity.

4.4.5 Active defense

Security countermeasures and protections can be applied passively or actively. Passive defenses include those defensive architecture techniques described in Section 4.4.2. These passive defenses establish barriers using defense-in-depth techniques to deter and protect against a malicious adversary. This technique, however, is static and reactionary. It is also at a disadvantage for defending against dynamic and adaptive adversary capabilities.

Instead of reliance on passive capabilities, engineers need to build in active defenses to preemptively prevent, detect, and respond to cyber incidents. This paradigm shift is needed to proactively identify malicious and inadvertent cyber incidents to quickly stop the incident and remove the threat before degradation or unrecoverable damage occurs. Active defenses include security information event monitoring and other real-time anomaly detection and response tools that may not yet be developed or deployed. The objective is to enhance resilience capabilities by improving operational situational awareness via dynamic and testable strategies. Ideally, active defense tools can identify cyber anomalies in all five threat vectors (e.g., wired networks, wireless networks, portable media and maintenance devices, insiders, and supply chain).

4.5 CIE Organizational principles

4.5.1 Interdependencies

The CIE organizational principles listed in **Figure 7** are those fundamental cybersecurity practices that enable holistic integration of cybersecurity into other programs within the facility or organization. Technical and administrative interdependencies are necessary for safe and secure reactor operation. From a technical perspective, this principle ensures that cybersecurity is considered within all the interconnections between systems and systems of systems, including extended data pathways. Additionally, 10CFR73.54 not only requires adequate protection of safety-related and important-to-safety SSCs but also those support systems relied upon to ensure safe operation of those functions. Support systems may include power, communications, water, or HVAC. Even though there is the potential for adverse safety or security consequences if a cyber incident impacts a support system, these interdependencies are often overlooked.

From an administrative perspective, the interdependency principle promotes a multidisciplinary approach to ensure all project personnel are involved. For instance, when designing or modifying a reactor safety system to perform specific functions, a design engineer relies on safety engineers to provide expertise on safety-related functions, quality engineers to verify correct design implementations, maintenance personnel to provide perspectives on accessibility and maintainability, operators to provide operational feedback under various conditions, and competent authorities to provide safety and security requirements.

With the shift towards DI&C, cyber engineers or specialists should also be included throughout the systems engineering lifecycle to provide valuable insight into cyber risk and risk (and cost) minimization strategies, such as cyber risk treatments, policies, and procedures. Additionally, it is paramount to ensure other disciplines, such as engineering, safety, risk, design, maintenance, operations, human factors, and ICT, are knowledgeable about these system interdependencies and the potential consequences of a cyber incident on a facility function, digital asset, system, or system of systems. While the nature of the multidisciplinary engagements may differ with each stage, similar to safety, the intent is to ensure cyber engineering remains a core domain throughout the entire lifecycle.

4.5.2 Digital asset inventory

Although new installations or modifications to existing facilities will include equipment database inventories of SSCs, this list often is out-of-date, incomplete, and without enough information to support cyber requirements and incident response decisions. Thus, the digital asset inventory CIE principle is intended to ensure that an accurate as-built digital asset inventory is maintained throughout the systems engineering lifecycle, including initial design, maintenance, configuration changes, and upgrades or modifications.

It is impossible to provide adequate protection against cyber incidents if there are unknown digital assets installed in a facility. Therefore, it is necessary to establish complete, accurate, and detailed asset inventories for the entire digital bill of materials (DBOM), including make, model, and version information for hardware, firmware, and software. For instance, if a vendor or intelligence agency provides vulnerability and threat information for a specific digital asset, a facility can easily use their inventory to determine if they have that asset installed. Accurate digital asset inventories improve the overall vulnerability management process. Without the inventory, it is very difficult to track whether newly identified cyber risks are applicable to the facility.

In addition to the DBOM, configuration information, backup requirements, and restoration information should be maintained for each digital SSC. Since cyber compromises do occur within the supply chain and early lifecycle stages, this complete design record should be maintained under secured configuration control such that all modifications or updates are captured. When used in conjunction with the incident response planning principle, this detailed information can be used to restore or rebuild a system after a cyber incident.

4.5.3 Supply chain and system information controls

The use of third-party digital hardware, firmware, and software has increased tremendously in the past several decades. The cost-benefit of purchasing general purpose multifunctional digital devices has become a mainstay for many custom in-house and engineered solutions. However, since vendors, integrators, and service providers are profit driven, they will likely not invest in additional cyber security designs and controls for their products and services unless required by procurement specifications.

Since the supply chain is one of five threat vectors into a nuclear facility, it is imperative to develop supply chain controls that incorporate techniques into the procurement and acquisition process to prevent malicious or inadvertent compromise of hardware, firmware, software, and system information, where system information is defined as the "complete record of information regarding a digital system or component, including system level and component level information and/or data such as requirements specifications, design documentation, fabrication, assembly or manufacturing details; validation and verification documentation; operation and maintenance manuals; credential, authentication, or cryptographic information; and product lifecycle plans" [39].

The primary objectives of cyber supply chain risk management include the ability to maintain authenticity, integrity, confidentiality, and exclusivity throughout the system engineering lifecycle [39]. Authenticity assures the components are genuine; integrity assures the components are trustworthy and uncompromised; confidentiality assures there is no unauthorized loss of data or secrets; and exclusivity assures there are limited touchpoints to reduce the number of attack points [40].

A simplified, notional DI&C supply chain cyber-attack surface is illustrated in **Figure 10**. It is important to understand this attack surface so appropriate risk treatments can be implemented to reduce cybersecurity risk throughout the lifecycle. Logically, the parallel use of the design simplification CIE principle reduces this supply chain cyber-attack surface by reducing the number of stakeholders and touchpoints. Ensuring cyber supply chain provenance and trustworthiness is easier with a smaller supply chain cyber-attack surface.

Procurement contracts should include cybersecurity requirements, such as those provided by the Department of Homeland Security [41], the Energy Sector Control Systems Working Group [42], or Electric Power Research Institute [43]. This procurement language should include all aspects of a product or service including the ability to review the supply chain stakeholder's cybersecurity program, including any assessments or cybersecurity testing. Without inclusion of cybersecurity requirements into procurement contracts, the likelihood of insecure or compromised products and services increases.

It is important to recognize that supply chain cybersecurity is necessary during early lifecycle stages even when only system information is available. Reconnaissance is a primary method used by an adversary to acquire preliminary information about an organization, operations, and system designs. Theft of confidential or proprietary system information may result in loss of intellectual property, counterfeiting, and enable development of future sophisticated cyber-attacks. In addition, compromise or falsification of system information could lead to developers inadvertently including malicious codes, falsified data, latent vulnerabilities, or backdoors into the system or component during supply chain activities.

Unfortunately, protection of sensitive information is historically inadequate—sensitive information can often be found on social media, corporate websites, conferences, business and employment-oriented online services, vendor advertising, and other third-party entities that store nuclear-related information, such as

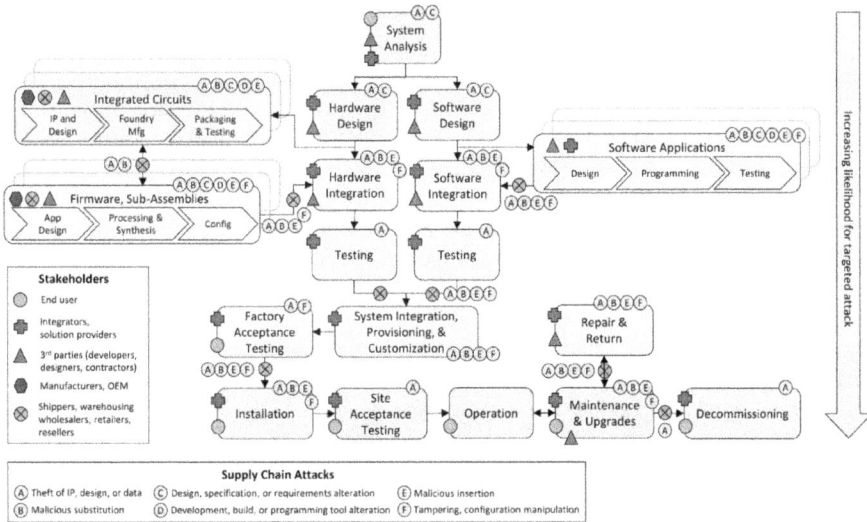

Figure 10.
A notional DI&C supply chain cyber-attack surface illustrating the complexity of the supply chain lifecycle overlaid with potential supply chain attacks at key stakeholder locations and touchpoints [39].

nuclear regulators. Of course, poor cybersecurity hygiene can occur at every stakeholder in the supply chain, including hardware manufacturers, programmers, and integrators, as well as the reactor designer and operator. Since engineering records, asset inventories, master drawings, procedures, specifications, analysis, and other sensitive system information is much more accessible today, responsibility for protecting system information lies not only with the entire nuclear organization but all supply chain stakeholders.

4.5.4 Incident response planning

Incident response planning, in conjunction with contingency planning in resilient design and an accurate and complete digital asset inventory, ensures that procedures, current backups, and accurate configurations are available to respond to and recover from deliberate or inadvertent cyber incidents. Cyber incidents can occur at any stage in the lifecycle. For example, theft of system information or IP can occur during design, introduction of malware by a subcontractor can occur during testing, and downloading of corrupted firmware can occur as part of maintenance. Incident response planning should occur in each stage of the systems engineering lifecycle to safeguard the stakeholder, system information, and DBOM against a cyber incident. IAEA TDL006 [44] and NIST 800-61 [45] provide incident response guidelines.

4.5.5 Cybersecurity culture and training

An organization's culture is demonstrated every day through the actions of its employees. Nuclear facilities are guided by a nuclear safety and security culture which emphasizes protection of public health and safety over other competing goals, such as electricity generation. Personnel are instilled with the understanding that they can and should speak up when there are safety or security concerns. Since cybersecurity is part of the overarching nuclear security policy to guard against theft and sabotage, developing and maintaining a cybersecurity culture and training program is just as important.

The human-in-the-loop is essential for maintaining a robust security posture. As digital technology is prevalent in both OT and ICT systems, every person is responsible for cybersecurity, not just ICT or engineering staff. Similar to the nuclear safety culture, an organization-wide cybersecurity culture and training program will equip all personnel with the knowledge, skills, and abilities to recognize, prevent, and respond to cyber incidents. The goal of CIE is to develop cyber-informed engineers and personnel as opposed to cybersecurity specialists. Development of cyber-awareness and cross-functional cyber capabilities will provide personnel with information on the importance of their role in an organization's overall security plan. Simply recognizing and reporting phishing emails or suspicious activity can prevent an adversary's entry into an organization. Without this knowledge of how cyber incidents can occur and what unauthorized interactions can look like, compromises can remain persistent and undetected, thereby leading to greater consequences for the organization or nuclear reactor.

5. Discussion

Applying the CIE approach throughout the entire systems engineering lifecycle, from design and testing to maintenance and decommissioning, provides enhanced capabilities for cyber protection, detection, and response. **Figure 11** is a notional diagram summarizing potential usage of CIE principles throughout the lifecycle. The primary objective of CIE is to ensure engineers and stakeholders consider CIE principles during each activity within every stage of the lifecycle. Continual cyber risk analysis ensures that new or updated consequences, threats, and vulnerabilities are quickly identified. CIE design principles ensure that approaches to address and reduce the identified cyber risk are considered to the greatest extent possible. And, finally, CIE organizational principles provide long-term cyber risk reduction benefits by holistically integrating cyber considerations throughout the facility and organization.

Since nuclear engineering projects differ in scope, it is impractical to define a standard level of effort for all CIE principles across each stage. For instance, the design and construction of an advanced reactor will likely have a very long timeline

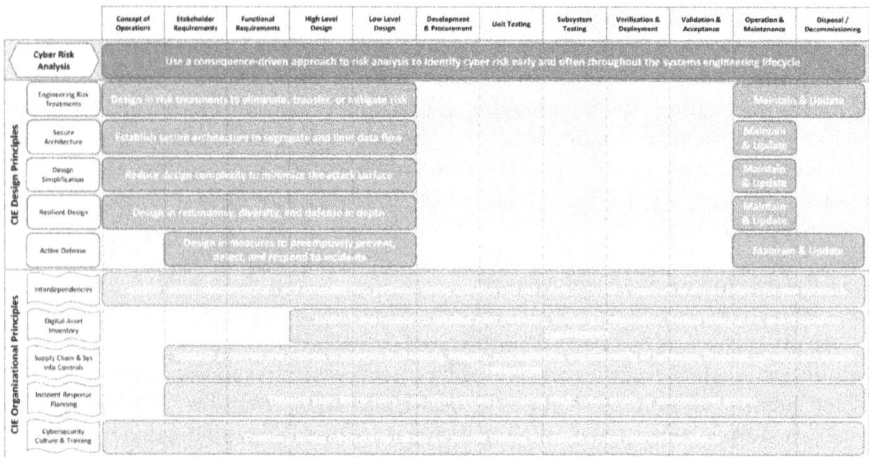

Figure 11.
Notional usage of CIE principles throughout the systems engineering lifecycle.

and involve multiple organizations, while a simple modification at a research reactor may occur relatively quickly and include only a small group of people. As an applied integrated energy system example, the CIE approach was used during the high-level design of a hydrogen generation project in which heat and electricity were provided by an interconnected NPP [35]. The use of a multi-disciplinary team to address system of system interdependencies through a structured risk analysis process resulted in new insights into the potential for both adversarial and unintentional cyber risks. As a result, the system was immediately redesigned to eliminate specific identified risk as well as to incorporate more simplified and resilient design features [35].

6. Conclusions

With the continued modernization of the existing nuclear fleet and future advanced reactor designs and applications, the use of DI&C in nuclear reactors will continue to grow. Additionally, once DI&C is installed or new reactors are commissioned, maintenance and updates will occur throughout a reactor's lifetime. The fundamental CIE objective to consider cyber requirements from the onset of conceptual design provides expanded opportunities for recognizing cyber risks, thereby enabling cyber risk reduction through redesign prior to initiation of any procurement or construction activities. While CIE can positively impact design modifications in existing reactors, it may have even greater potential in improving the security posture of new reactors. Convening multidisciplinary teams will enable novel cyber solutions that otherwise would not be possible, thus minimizing cybersecurity-related costs and expensive rework later in the lifecycle. Addressing cyber concerns after installation with bolt-on solutions is arguably less effective and less efficient, especially given the fact that some SSCs may not tolerate or allow the use of security controls.

CIE is a multidisciplinary approach incorporating design and organizational principles to protect digital technology from cyber risk. The continued adoption of CIE in nuclear organizations as well as the development of curriculum in academic engineering and industry education programs furthers the goal of globally reducing nuclear cyber risk.

Acknowledgements

The authors wish to acknowledge the contributions of Dr. Katya Le Blanc and Timothy R. McJunkin, who provided critical reviews and suggestions.

Funding

This work was supported by the U.S. DOE Office of Nuclear Energy Cybersecurity Crosscutting Technology Development program under the DOE Idaho Operations Office, Contract DE-AC07-05ID14517.

Conflict of interest

The authors declare no conflict of interest.

Acronyms

AI	artificial intelligence
CDA	critical digital asset
CCE	consequence-driven, cyber-informed engineering
CIE	cyber-informed engineering
DBA	design basis accident
DBOM	digital bill of material
DBT	design basis threat
DI&C	digital instrumentation and control
DOE	Department of Energy
ESFAS	engineered safety feature actuation system
GDC	general design criteria
HCE	high consequence event
I&C	instrumentation and control
IAEA	International Atomic Energy Agency
ICT	information and communications technology
IEEE	Institute of Electrical and Electronics Engineers
IIoT	industrial internet of things
IoT	internet of things
LWR	light water reactor
ML	machine learning
NEI	Nuclear Energy Institute
NIST	National Institute of Standards and Technology
NPP	nuclear power plant
NRC	Nuclear Regulatory Commission
NSS	nuclear security series
OT	operational technology
PRA	probabilistic risk analysis
RPS	reactor protection system
SMR	small modular reactor
SNM	special nuclear material
SSC	systems, structures, and components
USB	universal serial bus

Author details

Shannon Eggers[*] and Robert Anderson
Idaho National Laboratory, Idaho Falls, ID, USA

[*]Address all correspondence to: shannon.eggers@inl.gov

IntechOpen

References

[1] Stout TM, Williams TJ. Pioneering work in the field of computer process control. IEEE Annals of the History of Computing. 1995;**17**(1):6-18. DOI: 10.1109/85.366507

[2] NEI. NEI 10-04: Identifying systems and assets subject to the cyber security rule. Revision 2. Washington, DC: Nuclear Energy Institute; 2012. Available from: https://www.nrc.gov/docs/ML1218/ML12180A081.pdf

[3] IEEE. IEEE 308-1971—IEEE standard criteria for class 1E electric systems for nucler power generating stations. New York, NY: Institute of Electrical and Electronics Engineers; 1971. DOI: 10.1109/IEEESTD. 1971.6714366

[4] 10 C.F.R. § 50 Appendix A. Domestic Licensing of Production and Utiliziation Facilities. 2007. Available from: https://www.nrc.gov/reading-rm/doc-collections/cfr/part050/part050-appa.html

[5] IEEE. IEEE 279-1971—Criteria for safety systems for nuclear power generating stations. New York, NY: Institute of Electrical and Electronics Engineers; 1971. DOI: 10.1109/IEEESTD.2012.6125207

[6] IEEE. IEEE 603-1991—IEEE Standard Criteria for Safety Systems for Nuclear Power Generating Stations. New York, NY: Institute of Electrical and Electronics Engineers; 1991. DOI: 10.1109/IEEESTD.1991.101077

[7] NRC. Regulatory Guide 1.152. Revision 3. Criteria for use of computers in safety systems of nuclear power plants. Washington, DC: U.S. Nuclear Regulatory Commission; 2011. Available from: https://www.nrc.gov/docs/ML1028/ML102870022.pdf

[8] IEEE. IEEE 7-4.3.2-2003—IEEE standard for digital computers in safety systems of nuclear power generating stations. New York, NY: Institute of Electrical and Electronics Engineers; 2003. DOI: 10.1109/IEEESTD.2003.94419

[9] IEC. IEC 61513:2011. Nuclear Power Plants—Instrumentation and Control Important to Safety—General Requirements for Systems. Geneva, Switzerland: Rev 2.0. International Electrotechnical Commission; 2011. Available from: https://webstore.iec.ch/publication/5532

[10] IAEA. Specific Safety Requirements No. SSR-2/1. Safety of Nuclear Power Plants: Design (Rev 1). Vienna: International Atomic Energy Agency; 2016. Report No. STI/PUB/1534. Available from: http://www-pub.iaea.org/MTCD/Publications/PDF/Pub1534_web.pdf

[11] IAEA. Specific Safety Guide No. SSG-39. Design of Instrumentation and Control Systems for Nuclear Power Plants. Vienna: International Atomic Energy Agency; 2016. Available from: http://www-pub.iaea.org/MTCD/Publications/PDF/Pub1694_web.pdf

[12] Kaplan S, Garrick BJ. On the quantitative definition of risk. Risk Analysis. 1981;**1**(1):11-27, 1981. DOI: 10.1111/j.1539-6924.1981.tb01350.x

[13] Eggers S, Le Blanc K. Survey of cyber risk analysis techniques for use in the nuclear industry. Progress in Nuclear Energy. 2021;**140**:1. DOI: 10.1016/j.pnucene.2021.103908

[14] NRC. NRC Information Notice 2003-14: Potential vulnerability of plant computer network to worm infection. Washington, DC: U.S. Nuclear Regulatory Commission; 2003. Document No. IN200314. Available from: https://www.nrc.gov/reading-rm/doc-collections/gen-comm/info-notices/2003/in200314.pdf

[15] NRC. NRC Information Notice 2007-15: Effects of ethernet-based, non-safety related controls on the safe and continued operation of nuclear power stations. Washington, DC: U.S. Nuclear Regulatory Commission; 2007. Available from: http://www.nrc.gov/reading-rm/doc-collections/gen-comm/info-notices/2007/in200715.pdf

[16] Langner R. Stuxnet: Dissecting a cyberwarfare weapon. IEEE Security & Privacy. 2011;9(3):49-51. DOI: 10.1109/MSP.2011.67

[17] Krebs B. Cyber incident blamed for nuclear power plant shutdown. Washington Post. June 5, 2008. Available from: http://www.washingtonpost.com/wp-dyn/content/article/2008/06/05/AR2008060501958.html

[18] Graham M. Context threat intelligence—the Monju incident. New York, NY: Context Information Security; 2014. Available from: https://www.contextis.com/en/blog/context-threat-intelligence-the-monju-incident

[19] 10 C.F.R. § 73.54. Protection of Digital Computer and Communication Systems and Networks. 2009. Available from: https://www.nrc.gov/reading-rm/doc-collections/cfr/part073/part073-0054.html

[20] Common Vulnerabilities and Exposures (CVE). The MITRE Corporation. Available from: https://cve.mitre.org/ [Accessed: September 29, 2020]

[21] Common Weakness Enumeration (CWE). The MITRE Corporation. Available from: https://cwe.mitre.org/ [Accessed: September 29, 2020]

[22] Common Vulnerability Scoring System (CVSS). FiRST. Available from: https://www.first.org/cvss/ [Accessed: September 29, 2020]

[23] ICS-CERT Alerts. Cybersecurity and Infrastructure Security Agency. Available from: https://us-cert.cisa.gov/ics/alerts [Accessed: September 29, 2020]

[24] NRC. Regulatory Guide 5.71. Cyber security programs for nuclear facilities. Washington, DC: U.S. Nuclear Regulatory Commission; 2010. Available from: http://pbadupws.nrc.gov/docs/ML0903/ML090340159.pdf

[25] NEI. NEI 08-09: Cyber security plan for nuclear power reactors. Revision 6. Washington, DC: Nuclear Energy Institute; 2010. Available from: https://www.nrc.gov/docs/ML1011/ML101180437.pdf

[26] NEI. NEI 13-10: Cyber security control assessments. Revision 6. Washington, DC: Nuclear Energy Institute; 2017. Available from: https://www.nrc.gov/docs/ML1723/ML17234A615.pdf

[27] IAEA. Nuclear Security Series No. 13. Nuclear Security Recommendations on Physical Protection of Nuclear Material and Nuclear Facilities (INFCIRC/225/Revision 5). Vienna: International Atomic Energy Agency; 2011. Available from: http://www-pub.iaea.org/MTCD/Publications/PDF/Pub1481_web.pdf

[28] IAEA. Nuclear Security Series No. 17. Computer Security at Nuclear Facilities. Vienna: International Atomic Energy Agency; 2011. Available from: https://www.iaea.org/publications/8691/computer-security-at-nuclear-facilities

[29] IAEA. NSS 42-G. Computer Security for Nuclear Security. Vienna: International Atomic Energy Agency; 2021. Available from: http://www-pub.iaea.org/MTCD/Publications/PDF/PUB1918_web.pdf

[30] IAEA. NSS 33-T. Computer Security of Instrumentation and Control Systems at Nuclear Facilities. Vienna: International Atomic Energy Agency;

2018. Available from: http://www-pub. iaea.org/MTCD/Publications/PDF/ P1787_web.pdf

[31] IEC. IEC 62645:2019. Nuclear power plants—Instrumentation, control and electric power systems—Cybersecurity requirements. Geneva, Switzerland: International Electrotechnical Commission; 2019. Available from: https://webstore.iec.ch/ publication/32904

[32] IEC. IEC 62443-3-2. Security risk assessment and system design. Geneva, Switzerland: International Electrotechnical Commission; 2020. Available from: https://webstore.iec.ch/ publication/30727

[33] Bochman AA, Freeman S. Countering Cyber Sabotage: Introducing Consequence-Driven, Cyber-Informed Engineering (CCE). Boca Raton, FL: CRC Press; 2021. DOI: 10.4324/ 9780367491161

[34] Consequence-Driven Cyber-Informed Engineering. Idaho National Laboratory. Available from: https://inl. gov/cce/ [Accessed: November 8, 2021]

[35] Eggers S, Le Blanc K, Youngblood R, McJunkin T, Frick K, Wendt D, et al., editors. Cyber-Informed Engineering case study of an integrated hydrogen generation plant. ANS 12th Nuclear Plant Instrumentation, Control and Human-Machine Interface Technologies (NPIC&HMIT); 2021 June 13-16, 2021; Online Virtual Meeting: American Nuclear Society.

[36] Anderson RS, Benjamin J, Wright VL, Quinones L, Paz J. Cyber-Informed Engineering. Idaho Falls, ID: Idaho National Laboratory; 2017. DOI: 10.2172/1369373

[37] Stouffer K, Pillitteri V, Lightman S, Abrams M, Hahn A. SP 800-82. Revision 2: Guide to industrial control

systems (ICS) security. Gaithersburg, MD: National Institute of Standards and Technology; 2015. DOI: 10.6028/NIST. SP.800-82r2

[38] IAEA. NSS 17-T. Rev 1. Computer Security Tecniques for Nuclear Facilities. Vienna: International Atomic Energy Agency; 2021. Available from: http://www-pub.iaea.org/MTCD/ Publications/PDF/PUB1921_web.pdf

[39] Eggers S. A novel approach for analyzing the nuclear supply chain cyber-attack surface. Nuclear Engineering and Technology. 2021;**53**(3):879-887. DOI: 10.1016/j. net.2020.08.021

[40] Windelberg M. Objectives for managing cyber supply chain risk. International Journal of Critical Infrastructure Protection. 2016;**12**:4-11. DOI: 10.1016/j.ijcip.2015.11.003

[41] DHS. Cyber Security Procurement Language for Control Systems. Washington, DC: Department of Homeland Security; 2009. Available from: https://us-cert.cisa.gov/sites/ default/files/documents/Procurement_ Language_Rev4_100809_S508C.pdf

[42] ESCSWG. Cybersecurity procurement language for energy delivery systems. Washington DC: Energy Sector Control Systems Working Group; 2014. Available from: https:// www.energy.gov/sites/default/ files/2014/04/f15/CybersecProcurement Language-EnergyDeliverySystems_ 040714_fin.pdf

[43] EPRI. Cyber Security in the supply chain: Cyber security procurement methodology. Revision 2. Palo Alto, CA: Electric Power Research Institute; 2018. Document No. TR 3002012753

[44] IAEA. TDL005. Computer security incident response planning at nuclear facilities. International Atomic Energy Agency; 2016. Available from: http://

www-pub.iaea.org/MTCD/
Publications/PDF/TDL005web.pdf

[45] NIST. SP 800-61. Rev 2. Computer
security incident handling guide.
Gaithersburg, MD: National Institute of
Standards and Technology; 2012. DOI:
10.6028/NIST.SP.800-61r2

Reliability Analysis of Instrumentation and Control System: A Case Study of Nuclear Power Plant

Mohan Rao Mamdikar, Vinay Kumar and Pooja Singh

Abstract

Instrumentation and control system (I&Cs) plays a key role in nuclear power plants (NPP) whose failure may cause the major issue in a form of accidents, hazardous radiations, and environmental loss. That is why importantly ensure the reliability of such system in NPP. In this proposed method, we effectively analyze the reliability of the instrumentation and control system. An isolation condenser system of nuclear power plant is taken as a case study to show the analysis. The methodology includes the dynamic behavior of the system using Petri net. The proposed method is validated on operation data of NPP.

Keywords: reliability, control system, nuclear power plant, isolation condenser system

1. Introduction

Instrumentation and control system (I&C) plays a vital role in the field of the nuclear industry. Nowadays I&C systems are embedded into the nuclear power plant (NPP) operation and reliability. Each component of NPP, such as transformers, valves, circuit breakers, heat exchangers. is equipped with digital I&C system whose reliability plays a vital role to avoid any accidents. Because these components are safety-critical systems (SCS) whose failure may cause huge losses in the form of economic loss, human resource damage, and environmental loss. As instrumentation and control systems are the important and first layer of safety, reliability, and stability in the NPP [1] that is the reason, it is essential to ensure the reliability of such a safety system. With, the introduction of digital control systems in the last few decades where the reliability of digital I&C must not be degraded. Therefore, researchers are rigorously working to address the dependability of the system. The dependability includes reliability, safety analysis, performance, and availability attributes that are ultimately related to security. The model checking may be used to various issues, which can lead to spurious actuation of the I&C system [2]. The transformation from analog to digital I&C safety systems added new challenges for researchers as well as software developers to deliver correct software reliability [3]. Based on this software experts could take essential steps early in the design phase of software by avoiding failures in I&C of NPP. The cyberattack occurred in the I&C system in the Iranian Bushehr nuclear power plant, where configure was destroyed by malicious code [4]. Therefore, it is essential to I&C systems required having secure and reliable to avoid any kind of attacks causing major accidents. Many researchers have put efforts to address the reliability analysis on such systems using various techniques, such as fault tree analysis (FTA), reliability block diagram (RBD), Bayesian network, etc.

This work proposes the reliability analysis of instrumentation and control system (I&C) of NPP using stochastic Petri net (SPN).

The organization of this paper is as follows. In Section 2, our focus is on the related work of the proposed work. In Section 3, we discuss the background and mathematical fundamentals. In Section 4, proposes the framework of the proposed method. In Section 5, the case study of the proposed work. In Section 6, reliability analysis of the proposed work. In Section 7, the validation part is covered. In Section 8, the conclusion is made with future work.

2. Related work

Zeller et al. [5] proposed a combined approach of Markov chain and component fault tree to analyze the complex software-controlled system in the automotive domain. The authors have addressed safety and reliability in modular form. However, authors have missed to validate the result and failed to express reliability accuracy in percentage.

Nidhin et al. [6] presented a survey for understanding radiation effects in SRAM-based FPGAs for implementing I&C of NPP. Authors have found that for implementing NPP with I&C in SRAM-based FPGAs, the effect of radiation issue is a major concern. To reduce radiation-related issues some components, which have SRAM-based FPGAs, must keep outside of reactor containment building (RCB). However, the authors have failed to discuss the case study.

Jia et al. [7] proposed an approach for the identification of vulnerabilities present in elements that affect the reliability of digital instrumentation and control system (DI&C) software life cycle using Bayesian network. A reliability demonstration of safety-critical software (RDSS) integrates the claim-argument-evidence (CAE) and sensitivity to estimate the reliability of the system. However, there is a limitation with BN that has no time constraints and dynamic property. Authors have missed addressing the reliability with validation from the real-time dataset.

Rejzek and Hilbes [8] proposed system-theoretic process analysis (STPA) for design verification and risk analysis of digital I&C of NPP. This method is considered as a prominent approach for analysis of the I&C system theoretically as the authors claim. However, the authors are not very much sure, that method correct result.

Torkey et al. [9] proposed a reliability improvement framework of the digital reactor protection system by transforming reliability block diagram to Bayesian belief network (BBN). The proposed method gives the highest availability as a result and found some modules are riskier than others of I&C. However, authors claim that it gives the highest availability but missed to validate the result with real-time data.

Kumar et al. [3] proposed a framework for predicting the reliability of the safety-critical and control system using the Bayesian update methodology. The authors have validated the result with real-time data of 12 safety-critical control systems of NPP. However, the result obtained is purely based on the failure data, if failure data is unavailable then it is difficult to predict the reliability.

Mamdikar et al. [10] devise a framework for reliability analysis, performance analysis that maps unified modified language (UML) to Petri net. The proposed framework is validated with 32 safety-critical systems of NPP. However, Petri net has a state space explosion problem as a system grows gradually, so it is not a generalized approach.

Nayak et al. [11] proposed a methodology called assessment of passive system reliability (APSRA) is used to estimate the reliability of the passive isolation condenser system of the Indian advanced heavy water reactor (AHWR). In this methodology, reliability is estimated through PSA treatment using generic data of the component. A classical fault tree analysis is used to find the root cause of the critical parameter, which leads to failure. However, the authors have failed to validate the result.

Kumar et al. [12] proposed a safety analysis framework that maps UML into the state-space model as Petri net of the safety-critical system of NPP. In this methodology, the result is validated on 29 different safety-critical systems of NPP. However, the authors have used Petri net that has a state space explosion problem.

Tripathi et al. [13] proposed a noble methodology dynamic reliability analysis of the passive decay heat removal system of NPP using Petri net. The authors have validated the estimated reliability based on the data available using fault tree analysis. Most of the system does not have such type data, and then it is difficult to validate the result with missing failure data. Therefore, this methodology may not applicable for every safety-critical system of NPP.

Buzhinsky and Pakonen [14] proposed an automated symmetry breaking approach for checking failure tolerance of I&C system. With this method a fewer failure combination has to be checked. The complex structure paired with various specifications has to be checked under failure assumptions, which is the limitation of this work.

Singh et al. [15] proposed a system modeling strategy for design verification of I&C of nuclear power plant using Petri net and converting PN into Markov chain. In this approach, verification is validated on real-time data. However, Petri net has a state-space explosion problem, in such circumstances, it is difficult to handle complex systems, which is the limitation of the work.

Xi et al. [16] proposed a test strategy based on the random selection of logic path by which provides reliability estimation and is used for control system testing in digital control software systems in the NPP. However, the authors have not been addressed and validated the reliability evaluation.

Bao et al. [17] proposed hazard analysis for identifying common cause failure of digital I&C using redundancy guided system in NPP. To conduct using redundancy guided systems, theoretic hazard analysis a modularized approach was applied. This method is helpful to remove casual effects of potential single points of failure that exist in I&C. However, authors have missed addressing the reliability analysis using this methodology in NPP.

Gupta et al. [1] proposed a method for stability analysis and steady-state analysis of the safety system of NPP using Petri net. The stability and steady-state were estimated and validated, however, authors have missed estimating reliability. The authors have to correlate stability with reliability. Further, this methodology is applicable only for discrete-time systems.

3. Background and mathematical fundamentals

This section consists of background and mathematical fundamentals to carry out reliability analysis of instrumentation and control system: a case study of nuclear power plant.

3.1 Petri net

A Petri net (PN) is mathematically defined 5-tuple $PN = (P, T, F, W, M_0)$ where P the finite is set places, T is a finite set of transitions, F is a finite set of arcs also referred to as flow relation, i.e., $F \ (P \times T) \ (T \times P)$, $W : F \ \{1, 2, 3, ...\}$ is the weight function, and M_0 is the initial marking $M_0 : P \{0, 1, 2, 3, ...\}$. $P\ T =$ and $P\ T \neq$. If the Petri net does not have an initial marking, it is denoted as $N = (P, T, F, W)$ with an initial marking denoted by (N, M_0). A simple example of the PN is shown in **Figure 1**.

The marking changes in the Petri net as per the transition firing are as follows:

  i. A transition in the enable mode when each input place of p of t is marked with at least $w(p, t)$ tokens.

  ii. An enabled transition is not necessarily fired.

  iii. A firing of enabled transition removes tokens from the input place and deposited in the output place.

3.2 Stochastic Petri net

A stochastic Petri net (SPN) is the extension of Petri net. In SPN, each transition is associated with a time delay that is an exponentially distributed random variable that expresses delay denoted by $SPN = (P, T, F, W, M_0)$.

3.3 Reachability

Reachability is the fundamental study of the dynamic property of the system. A marking M_n is said to be reachable from another marking M_1 if there exists a firing sequence that transforms M_n to M_1 such that $\partial = \{M_1 t_0 M_2 t_1 M_3 ... . t_n M_n\}$.

3.4 Reachability graph and Markov chain (MC)

A marking M is reachable from the initial marking M_0 if there exists a firing ∂ that brings back from the initial state of PN to a state that corresponds to M_0.

The Markov chain (MC) is the Markov process with discrete state space. The MC is obtained from the reachability graph of the SPN. Let SPN be the reversible, i.e., $M_0 \in R(M_i)$ for every M_i in $R(M_0)$, then the SPN generates an ergodic continuous time Markov chain (CTMC) and it is possible to compute the steady-state probability distribution $\prod$ by solving the following (Eq. (1)) and (Eq. (2)).

$$\sum \prod Q = 1 \qquad (1)$$

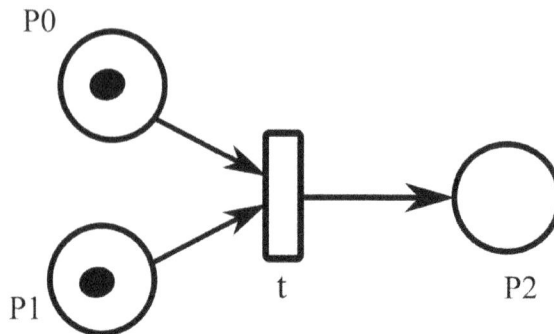

Figure 1.
Simple Petri net.

$$\sum_{i=1}^{s} \pi_i = 1 \tag{2}$$

Where, π_i is the probability being in the state M_i and $\prod = (\pi_1, \pi_2, \ldots \pi_s)$.

4. Framework of the proposed method

The proposed framework has six steps shown in **Figure 2**. Step 1—based on the system requirement we model the stochastic Petri net.

In step 2—by executing the PN model, we generate possible tangible states. Based on the tangible states, we construct the reachability graph in step 3. In step 4—obtained Markov chain form reachability, the graph of SPN. In step 5, we estimate the reliability of the ISO system. In step 6, we validate the result with real-time operation data of NPP.

5. Case study: Isolation condenser system (ISO)

The isolation condenser system simply referred to as ISO is a standby high-pressure system that removes residual and decay heat from the reactor vessel in the event of a scram signal in which the reactor becomes isolated from the main condenser, or if any other high-pressure condition exists. The schematic diagram is shown in **Figure 3**. The ISO system transfers residual and decays heat from the reactor coolant to the water in the shell side of the isolation condenser resulting in steam generation (SG). The steam generated in the shell side of the isolation condenser is then vented to the outside atmosphere. During the normal operation, the ISO system is in standby mode. During the standby mode, the steam isolation valves (VS1 and VS2) are open because the condenser tube bundles are at the reactor

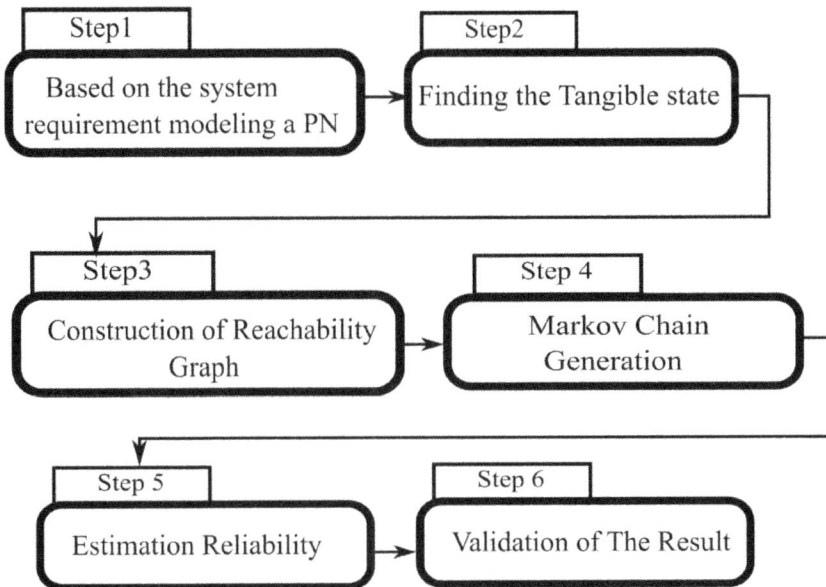

Figure 2.
Proposed framework of the system.

pressure. The condensate is built in the condenser and condensate by returning pipe. The condensate is stopped from a return back to the reactor by closing the condensate return valve (V_{C2}). The condensate valve (V_{C1}) is open at the stand-by condition and vent valves (VV) at main steam lines normally open to vent noncondensable gases from ISO. The makeup water must be provided to prevent uncovering the condenser tubes that are the combination of firewater and condensate using makeup water valve (V_W) normally closed at standby mode. The water inventory on the shell side of the condenser will provide heat removal for between 20 and 90 minutes depending on the plant design, at which time makeup water must be provided to prevent uncovering the condenser tubes. On the shell side of the condenser, the water inventory will be provided for the heat removal between 20 to 90 minutes. At which time water makeup has to be provided to prevent uncovering the condenser system tubes (**Figure 3**).

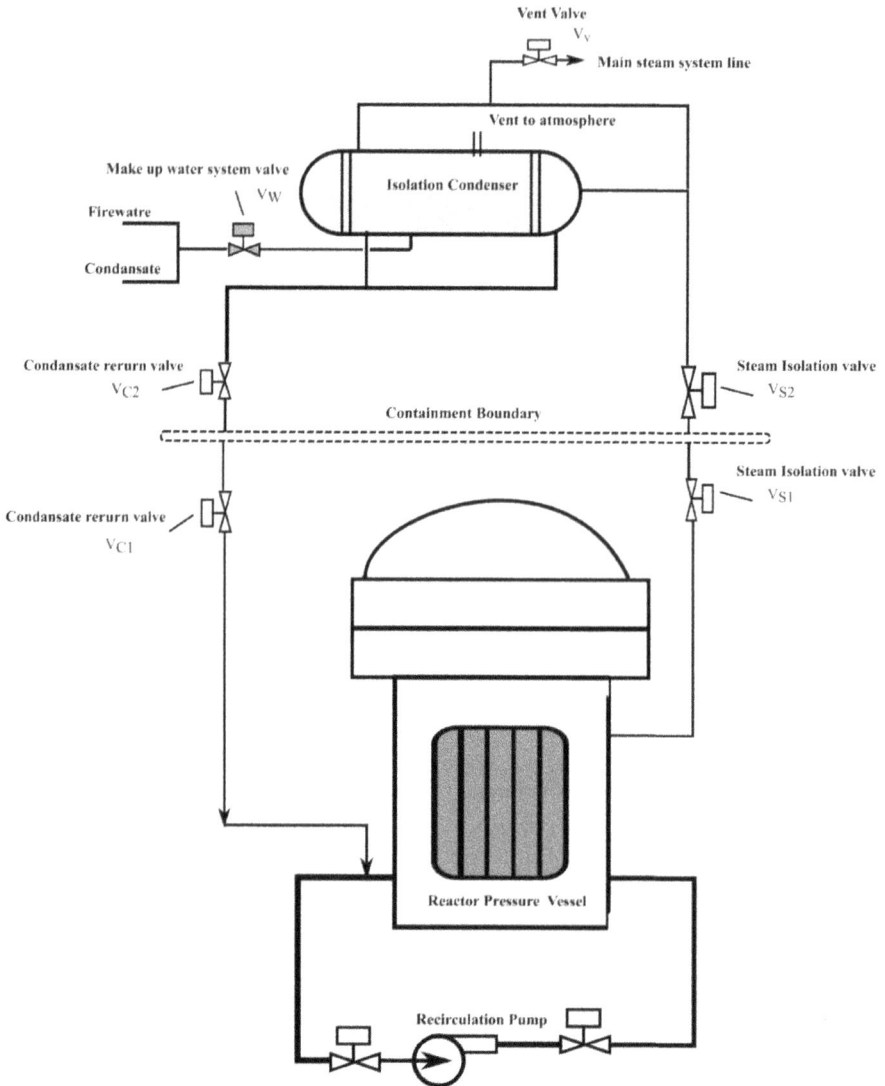

Figure 3.
Schematic diagram of isolation condenser system.

The ISO system may be initiated manually, or automatically initiated on high reactor pressure or low reactor pressure. On the initiation of ISO, one of the condensate return valves (V_{C2}) opens and the vent valve (V_V) gets closed. The steam flows from the reactor vessel to steam isolation valves (V_{S1} and V_{S2}). The steam gets condensed in condenser tube bundles and condensed steam returns to the reactor vessel (V_{C2} and V_{C2}) with help of a recirculation pump. The boiled-off water is replaced by the condensate transfer system or the firewater system. The ISO system is designed in such a way that, the system automatically gets isolated from the reactor pressure vessel in the event of a system pipe break. All the valves are closed automatically (V_{S1}, V_{S2}, V_{C2}, V_{C2}, and V_V) in the event of low differential pressure exceeds three times the normal flow value. This isolation will mitigate the loss of water inventory. The ISO system instrumentation and control consists of initiation and containment isolation circuitry [18]. These circuits provide different functions, both of which are important to system reliability. The entire system is operating in a closed-loop manner.

6. Proposed framework of approach

To estimate the reliability by our approach of the ISO which consist of six steps as shown in **Figure 2** as described step by step as follows:

6.1 PN model generation

In this phase, we construct the PN model of ISO system based on system requirements and specifications. As several researchers have proposed methods [19], based on that we generated a PN model. Based on functional requirements, the activity involves the PN generation to identify the places and transitions of the case study: ISO system. The identified places and transitions as illustrated in **Table 1**.

Thereafter, we use the TimeNet4.5 [20] tool for SPN creation. Then we assign the transition delay to the transition based on the system requirement. To get

Places	Description	Transitions	Description
P0	Sensors detect trip	T0	Sensors detects initial condition
P1	Initial signal generated	T1	Triggers V_V valve close and V_V valve close
P2	Initial condition holds	T2	IC loop triggers
P3	Initial condition forwards	T3	Triggers V_V valve open and V_V valve close
P4	IC loop activated	T4	Triggers V_{s1} valve and V_{s2} valve open
P5	V_{c2} valve close	T5	Send signal to V_w valve open
P6	V_{c1} valve open	T6	Triggers V_w valve open
P7	V_V valve close	T7	Reset
P8	V_{s1} valve open	T8	Reset of AC loss
P9	V_{s2} valve open	T9	Reset of restoration
P10	Level measure makeup		
P11	V_w valve open		
P12	Reset		

Table 1.
ISO places and transitions based on function specification.

throughput values of transition stationary analysis was performed in the TimeNet tool as shown in **Table 2**.

The PN model was generated using TimeNet tools shown in **Figure 4**.

6.2 Tangible states and reachability graph creation

Tangible states are those for timed transitions [21], since we used SPN so there are e tangible states with markings as shown in **Table 3**.

Based on the tangible states of the PN a reachability graph of the PN (**Figure 4**) can be obtained as shown in **Figure 5**.

Transition	Rate	Symbol	Throughput value
T0	1 ms	$\lambda 0$	0.26966908
T1	1 ms	$\lambda 1$	0.10385724
T2	1 ms	$\lambda 2$	0.28610826
T3	1 ms	$\lambda 3$	0.1771261
T4	1 ms	$\lambda 4$	0.08883328
T5	1 ms	$\lambda 5$	0.09000000
T6	1 ms	$\lambda 6$	0.03244971
T7	1 ms	$\lambda 7$	0.06681974
T8	1 ms	$\lambda 8$	0.03152016
T9	1 ms	$\lambda 9$	0.03244971

Table 2.
ISO throughput values.

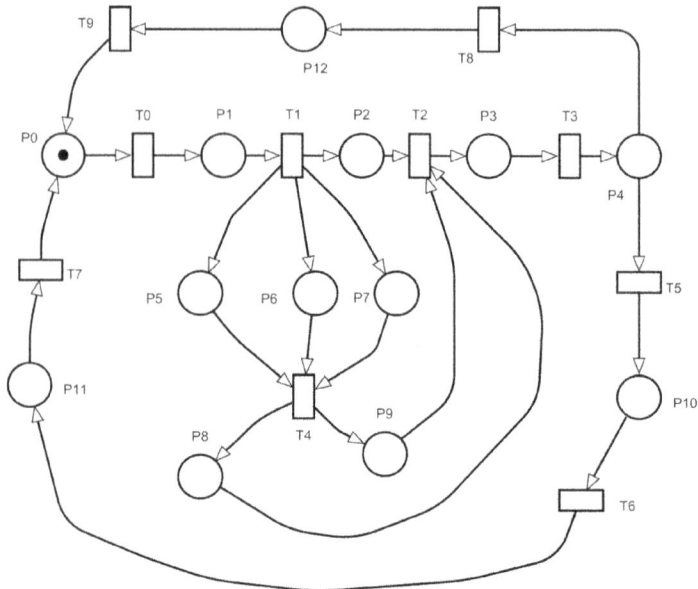

Figure 4.
PN model of ISO.

States	Marking	Tangible
M_0	1,000,000,000,000	Yes
M_1	0100000000000	Yes
M_2	0000010011100	Yes
M_3	0000010000011	Yes
M_4	0000001000000	Yes
M_5	0000000100000	Yes
M_6	0000100000000	Yes
M_7	0010000000000	Yes
M_8	0001000000000	Yes

Table 3.
ISO tangible states with markings of PN.

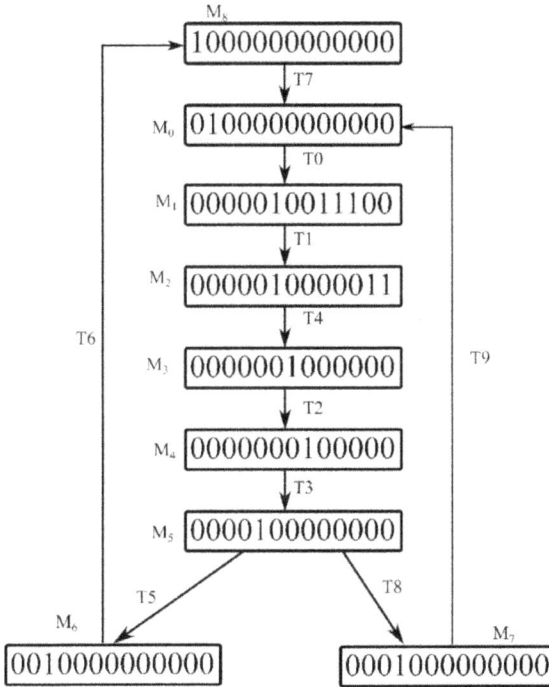

Figure 5.
Reachability graph.

6.3 Markov chain model creation

The MC model shown in **Figure 6** is obtained from the reachability graph of the PN shown in **Figure 4**.

With the help of Q which is transition probability matrix, the transition probability P_{ij} of MC can be computed from SPN. For the transition matrix Q, transitionrate q_{ij} is the transition of one state to another states unit/per time, therefore we take the ratio of the transition q_{ij} and the transition rate of the states sum must be zero. The diagonal elements can be defined as:

$$q_{ii} = -\sum_{j \neq i} q_{ij} \tag{3}$$

It is clear that the system is no ergodic, therefore, P_{ij} will be zero and defined as:

$$P_{ij} = \begin{cases} \dfrac{q_{ij}}{\sum_{k \neq i} q_{ik}}, & \text{if } k \neq i \\ 0, & \text{otherwise} \end{cases} \tag{4}$$

$P = I - d_Q^{-1}Q$, where $d_Q = dia(Q)$ diagonal matrix of Q.
The transition matrix is given in Eq. (5) as follows:

	M_0	M_1	M_2	M_3	M_4	M_5	M_6	M_7	M_8
M_0	$-λ0$	$λ0$	0	0	0	0	0	0	0
M_1	0	$-λ1$	$λ1$	0	0	0	0	0	0
M_2	0	0	$-λ4$	$λ4$	0	0	0	0	0
M_3	0	0	0	$-λ2$	$λ2$	0	0	0	0
M_4	0	0	0	0	$-λ3$	$λ3$	0	0	0
M_5	0	0	0	0	0	$-(λ5 + λ8)$	$λ8$	$λ5$	0
M_6	$λ9$	0	0	0	0	0	$-λ9$	0	0
M_7	0	0	0	0	0	0	0	$-λ6$	$λ6$
M_8	$λ7$	0	0	0	0	0	0	0	$-λ7$

$=$

	M_0	M_1	M_2	M_3	M_4	M_5	M_6	M_7	M_8
M_0	-0.26966	0.26966	0	0	0	0	0	0	0
M_1	0	-0.1038	0.1038	0	0	0	0	0	0
M_2	0	0	-0.0888	0.0888	0	0	0	0	0
M_3	0	0	0	-0.2861	0.2861	0	0	0	0
M_4	0	0	0	0	-0.1771	0.1771	0	0	0
M_5	0	0	0	0	0	-0.0630	0.0315	0.090	0
M_6	0.0324	0	0	0	0	0	-0.0324	0	0
M_7	0	0	0	0	0	0	0	-0.0324	0.0324
M_8	0.0668	0	0	0	0	0	0	0	-0.0668

$$\tag{5}$$

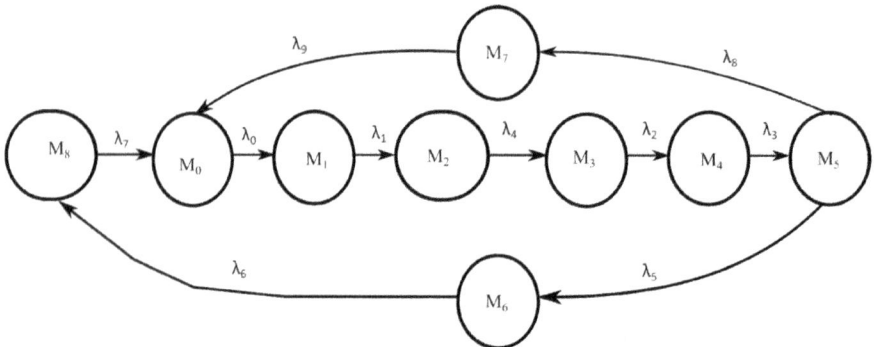

Figure 6.
Markov chain.

Now we solve Eq. (5) to get the design metrics and it seriousness of the NPP as defined in Eq. (6). We solve the Eq. (6) then we get the following linear equations.

$$
\begin{array}{c|ccccccccc}
 & M_0 & M_1 & M_2 & M_3 & M_4 & M_5 & M_6 & M_7 & M_8 \\
\hline
M_0 & 0 & 1 & 0 & 0 & 0 & 0 & 0 & 0 & 0 \\
M_1 & 0 & 0 & 1 & 0 & 0 & 0 & 0 & 0 & 0 \\
M_2 & 0 & 0 & 0 & 1 & 0 & 0 & 0 & 0 & 0 \\
M_3 & 0 & 0 & 0 & 0 & 1 & 0 & 0 & 0 & 0 \\
M_4 & 0 & 0 & 0 & 0 & 0 & 1 & 0 & 0 & 0 \\
M_5 & 0 & 0 & 0 & 0 & 0 & 0 & 0.2 & 0.7 & 0 \\
M_6 & 1 & 0 & 0 & 0 & 0 & 0 & 0 & 0 & 0 \\
M_7 & 0 & 0 & 0 & 0 & 0 & 0 & 0 & 0 & 1 \\
M_8 & 1 & 0 & 0 & 0 & 0 & 0 & 0 & 0 & 0
\end{array}
\tag{6}
$$

$$M_0 = M_1 \tag{7}$$

$$M_2 = M_1 \tag{8}$$

$$M_2 = M_3 \tag{9}$$

$$M_3 = M_4 \tag{10}$$

$$M_4 = M_5 \tag{11}$$

$$M_5 = 0.2M_6 \tag{12}$$

$$M_5 = 0.7M_7 \tag{13}$$

$$M_6 = M_0 \tag{14}$$

$$M_7 = M_8 \tag{15}$$

$$M_8 = M_0 \tag{16}$$

$$\sum_{i=0}^{8} M_i = 1 \tag{17}$$

6.4 Reliability analysis of proposed framework

Let $p_i(t)$ be the probability which component in state at time t is i. When components execute for $t \to \infty$ then probability leads to the stationary distribution. Then probability is defined as:

$$\vec{p}\,[p(M_0), p(M_1), p(M_2), p(M_3), p(M_4), p(M_5), p(M_6), p(M_7), p(M_8),] \tag{18}$$

$$\sum_{i \in M} p(i) = 1 \tag{19}$$

$$Reli_{ISO}^{est} = 1 - \sum_{i \in 6} M_i \tag{20}$$

There is only one failure state M_6 in MC. Now we solve the linear equation Eqs. (7)-(16) and Eq. (17) using the standard method, we get steady-state probability of each state as follows:

$M_0 = 0.1282051$, $M_1 = 0.1282051$, $M_2 = 0.1282051$, $M_3 = 0.1282051$, $M_4 = 0.1282051$, $M_5 = 0.1282051$, $M_6 = 0.025641$, $M_7 = 0.1025641$, and $M_8 = 0.1025641$

Hence the reliability of ISO is:

$$\text{Reliability} = 1 - 0.025641 = 0.974359. \tag{21}$$

7. Validation of proposed framework

In section, we compute the rate of failure to ensure the result experimentally of the proposed framework and follow the six steps for reliability estimation [10, 22]. We divide the entire input class into several subclass and for estimating reliability following equation required as:

$$R(t) = \sum_{i=1}^{6} \frac{h_i}{n_i} P(e_i) \tag{22}$$

$P(e_i)$ is the probability specified from input operation data. n_i is the number of trials from each comparable class. h_i is a number of trial cases that are failed.

To estimate the actual reliability **Table 4** data will be used.

Now using Eq. (22) we estimate actual reliability as:

$$Rel_{actual} = 1 - \sum_{i=1}^{6} \frac{h_i}{n_i} P(e_i) = 0.989999$$

Now we compare estimated (predicted) and actual reliability as:

$$Reli(diff) = Rel_{actual} - Reli_{estimated}$$

$$= 0.989999 - 0.974359$$

$$= 0.01564$$

Hence, the error percentages can be computed as:

$$Error\% = \frac{Rel(diff)}{Rel_{actual}} X100 = \frac{0.01564}{0.989999} X100 = 1.57981\%$$

Hence, the accuracy of proposed reliability computed of proposed framework is $(100 - error\%) = 98.4201\%$ that indicates the validation of our work.

Class	$P(e_i)$	h_i	n_i	$\frac{h_i}{n_i}P(e_i)$
Triggers V_V valve close and V_V valve close	0:028	2	170	0.00039
IC loop triggers	0.023	1	200	0.000115
Triggers V_V valve open and V_V valve close	0.0304	4	200	0.000608
Triggers V_{s1} valve and V_{s2} valve open	0.0987	2	40	0.004935
Send signal to V_w valve open	0.0342	3	30	0.00342
Triggers V_w valve open	0.0032	5	30	0.000533

Table 4.
Reliability estimation using [22].

8. Conclusion

The proposed method is centered technique for computing reliability of instrumentation and control system of the safety-critical system of NPP. We have validated the result with operational and found accuracy with 98.4201%. With this method, software designers take necessary preventive measures early design phase to avoid any kind of failure.

Author details

Mohan Rao Mamdikar[1], Vinay Kumar[1*] and Pooja Singh[2]

1 National Institute of Technology, Jamshedpur, India

2 VJIT, Mumbai, India

*Address all correspondence to: vkumar.cse@nitjsr.ac.in

IntechOpen

References

[1] Gupta B, Singh P, Singh L. Stability and steady state analysis of control and safety systems of nuclear power plants. Annals of Nuclear Energy. 2020;**147**:107676

[2] Pakonen A, Buzhinsky I, Björkman K. Model checking reveals design issues leading to spurious actuation of nuclear instrumentation and control systems. Reliability Engineering and System Safety. 2021; **205**:107237

[3] Kumar P, Singh LK, Kumar C. Software reliability analysis for safety-critical and control systems. Quality and Reliability Engineering International. 2020;**36**(1):340-353

[4] Chung M, Ahn W, Min B, Seo J, Moon J. An analytical method for developing appropriate protection profiles of instrumentation & control system for nuclear power plants. The Journal of Supercomputing. 2018;**74**(5): 1-16

[5] Zeller M, Montrone F. Combination of component fault trees and Markov chains to analyze complex, software-controlled systems. In: 2018 3rd International Conference on System Reliability and Safety (ICSRS). 23-25 November 2018; 2019. pp. 13-20

[6] Nidhin TS, Bhattacharyya A, Behera RP, Jayanthi T, Velusamy K. Understanding radiation effects in SRAM-based FPGAs for implementing instrumentation and control systems of nuclear power plants. Nuclear Engineering and Technology. 2017;**49**: 1589-1599

[7] Jia G, Ming Y, Bowen Z, Yuxin Z, Jun Y, Xinyu D. Annals of nuclear energy nuclear safety-critical digital instrumentation and control system software: Reliability demonstration. Annals of Nuclear Energy. 2018;**120**: 516-527

[8] Rejzek M, Hilbes C. Use of STPA as a diverse analysis method for optimization and design verification of digital instrumentation and control systems in nuclear power plants. Nuclear Engineering and Design. 2018;**331**:125-135

[9] Torkey H, Saber AS, Shaat MK, El-Sayed A, Shouman MA. Bayesian belief-based model for reliability improvement of the digital reactor protection system. Nuclear Science and Techniques. 2020; **31**(10):1-19

[10] Mamdikar MR, Kumar V, Singh P, Singh L. Reliability and performance analysis of safety-critical system using transformation of UML into state space models. Annals of Nuclear Energy. 2020;**146**:107628

[11] Nayak AK et al. Reliability assessment of passive isolation condenser system of AHWR using APSRA methodology. Reliability Engineering & System Safety. 2009;**94**: 1064-1075

[12] Kumar V, Singh LK, Singh P, Singh KV, Maurya AK, Tripathi AK. Parameter estimation for quantitative dependability analysis of safety-critical and control systems of NPP. IEEE Transactions on Nuclear Science. 2018; **65**(5):1080-1090

[13] Tripathi AM, Singh BLK, Singh CS. Dynamic reliability analysis framework for passive safety systems of nuclear power plant. Annals of Nuclear Energy. 2020;**140**:107139

[14] Buzhinsky I, Pakonen A. Symmetry breaking in model checking of fault-tolerant nuclear instrumentation and control systems. IEEE Access. 2020;**8**: 197684-197694

[15] Singh LK, Vinod G, Tripathi AK. Design verification of instrumentation and control systems of nuclear power

plants. IEEE Transactions on Nuclear Science. 2014;**61**(2):921-930

[16] Xi W, Liu W, Bai T, Ye W, Shi J. An automation test strategy based on real platform for digital control system software in nuclear power plant. Energy Reports. 2020;**6**:580-587

[17] Bao H, Shorthill T, Zhang H. Hazard analysis for identifying common cause failures of digital safety systems using a redundancy-guided systems-theoretic approach. Annals of Nuclear Energy. 2020;**148**:107686

[18] Kvarfordt KJ, Schroeder JA, Wierman TE. System study: Isolation condenser 1998–2018. December 2019

[19] Murata T. Petri nets: Properties, analysis and applications. Proceedings of the IEEE. 2015;**77**(4):541-580

[20] Zimmermann A, German R. Petri Net Modelling and Performability Evaluation with TimeNET 3.0. 2000. pp. 188-202

[21] Akharware N, Miee M, editors. PIPE2: Platform Independent Petri Net Editor. 2005

[22] Brown JR, Lipow M. Testing for software reliability. ACM SIGPLAN Notices. 1975;**10**(6):518-527

Chapter 8

Plenum Gas Effect on Fuel Temperature

Alok Jha

Abstract

All key phenomena in a fuel element are dominated by the temperature distribution. Fuel thermal expansion, fission gas-induced swelling, and release are directly related to the temperature distribution of the fuel. The fuel-cladding heat transfer coefficient has two components (a) heat transfer through the plenum and (b) heat transfer in case of contact. The gap width, in turn, is affected by thermal expansion, cracking and healing of the fuel, fuel densification, and fuel swelling. As the thermal and mechanical properties of the fuel are interdependent, inaccuracy in fuel-cladding temperature difference directly affects the reactor operating margins. A quantitative, as well as qualitative assessment of the fission heat transport across the fuel and embodiment of that knowledge in computer code, allows for a more realistic prediction of fuel performance. This knowledge helps in reducing the operating margins and leads to an improved operating economy of the reactor.

Keywords: fuel plenum, fuel thermal properties, fuel mechanical properties, computer code, reactor operation, thermal conductivity

1. Introduction

A good understanding of the factors governing the temperature distribution within a nuclear fuel element is important to predict the fuel temperature in all operating conditions of a nuclear reactor. The temperature distribution influences fuel performance in terms of solid-state reactions, e.g., grain-growth, densification, etc., and the temperature gradient results in fuel deformation or crack in low temperature zones. Oxide fuel is particularly disadvantageous due to its low density and low thermal conductivity [1]. Hence, a large temperature difference between the center and the surface of the rod is required for efficient heat extraction to make electric power generation economical. These constraints are at odds with each other. We intend to operate the reactor at the largest possible power density consistent with maintaining the fuel and coolant temperature below limits set by safety considerations. In accident conditions, we need to have enough margin so that the fuel does not lose integrity due to high temperature and poor heat transfer arrangement. Hence, the length of time and the fuel element that can be utilized in the reactor core is determined by the ability of the fuel element to withstand radiation damage and thermal and mechanical stresses experienced in the reactor environment and not so much on the depletion of fissile material. This is true for reactors utilizing enriched uranium as well as those using natural uranium as fissile fuel material.

Uranium metal is superior to oxide as far as density and conductivity is concerned, but the phase change at a low temperature of 600°C followed by a large volume change means that the fuel clad will be under severe stress. This has led to the investigation of other refractory compounds of uranium, such as uranium carbide, uranium nitride. For any type of fuel being used in the reactor, the fuel performance computer codes are needed to assure the continued safe operation of the reactor. With increasing demands of nuclear fuel efficiency, new fuel designs are being studied and the reliability of these new designs is in the interest of fuel manufacturers [1].

In this chapter, we will look into existing fuel analysis computer codes to develop an appreciation of the fuel characteristics. In the last section of this chapter, we will discuss a new code Fuel Characteristics Calculator (FCCAL) [2] and its suitability in the analysis of oxide fuel.

2. Classification of fuel performance codes

There are several available computer codes to analyze the thermal and mechanical behavior of fuel for different types of reactors viz., LWR, CANDU, VVER, etc. some of these codes are available in the public domain while some are proprietary and not available publicly.

The fuel rod behavior is determined by thermal, mechanical, and physical processes such as densification, swelling, fission gas generation, fission gas release, and irradiation damage. The fuel performance analysis code covers these aspects through thermal and mechanical components of fuel performance. The codes may be 1D, 2D, or 3D. However, experience shows that one-dimensional codes are most widely used for fuel analysis. The codes can be further classified as steady-state and/or transient codes. Examples of steady-state codes are FRAPCON, TRANSURANUS, COMETHE, etc. these codes calculate the radial temperature profile and fission gas release to the fuel plenum. Mechanical properties like creep deformation and irradiation growth can also be calculated using these codes. The transient codes like GRASS-SST [3] can calculate these parameters and additionally calculate cladding plastic stress-strain behavior, the effect of annealing, the behavior of oxide and hydrides during temperature ramps, phase changes, and large cladding deformation such as ballooning. The transient codes neglect long-term phenomena like creep deformation. Let us first discuss two computer codes for an understating of how we go about fuel characteristics quantification.

2.1 GAPCON-THERMAL

GAPCON-THERMAL-II (GT-II) [4] is an updated version of the older GAPCON-THERMAL-I (GT-I) [5] code that is widely used for calculating light water reactor fuel thermal performance. GT-I has been modified to improve upon the uncertainty in the calculation of power history and burn up. GT-II is an American National Standards Institute (ANSI) compliant Fortran-77 code. We can calculate the thermal behavior, fuel plenum conductance, temperature and pressure, and fuel stored energy using this code. There are models for power history, fission gas generation and release, fuel relocation, and densification in the code. For the power history simulation, the code uses constant power for each finite time step. At any time step other than the first for each axial node, the current fission gas release, relocation, and densification values are compared with the values used in the previous step. Relocation and densification displacement will not decrease if lesser values are subsequently calculated. The fission gas release algorithm depends upon all previous fission gas release values. The fission gas generation does not require that the simulation starts from zero. Transmutation of U-238 to plutonium and subsequent fission gas release due to fission of plutonium is also available in the code models. The fuel diameter is a function of power (Kw/ft), as fabricated cold-gap thickness (inches) and burnup (MWD/MTM). The model is based on linear regression analysis of experimental data. The fuel densification correlation calculates the reduction of fuel radius as a function of burnup and differential fuel density.

GT-II calculates the gap conductance, temperature, pressure, and stored thermal energy based on the power history of the fuel. The plenum gap conductance for each equal-length, user-designated axial region is determined by an iterative scheme. Radial temperature is calculated using finite difference. The solution procedure consists of iterative convergence for each axial region, followed by iterative convergence on the fuel gas release for each time-power step. Empirical, theoretical, and physical models are used for fuel gas release calculation.

2.2 Fuel design analysis (FUDA)

The computer code FUDA [6] is used for the design analysis of fuel for licensing application of CANDU type reactors in India. The code is used for fuel performance evaluation as well as to optimize the fuel design and fabrication parameters of Indian reactor fuel. The code is valid for the burnup of 50,000 MWD/Te of oxide fuel. Natural uranium and thorium oxide fuel can be analyzed using this code. There are models for computation of fuel temperature, thermal expansion, and clad stress parameters in the code.

The code uses the finite difference method for temperature and computation of thermal expansion. The clad stress, local stress, and ridge analysis is carried out by finite element technique. Fuel expansion is calculated by the two-zone model in which the stress in uranium oxide is ignored. Uranium oxide deformation is assumed to occur above a certain temperature as plastic, and below this temperature, the fuel element is assumed to behave as elastic solid with radial cracking. The extent of plasticity is governed by fuel temperature, stress due to cladding strength, and the coolant pressure in a time-dependent manner. Global clad stress and strain due to fuel thermal expansion, swelling and densification are calculated by models and correlation used in Notley [7]. The creep and stress relaxation in the time zone at constant power operation is calculated using semi-empirical formula considering a thermal and thermal creep including the effect of irradiation. Fuel sheath interfacial pressure is then calculated based on gas pressure and strains. Using global diametral changes, local deformation of the fuel element and sheath is calculated considering hourglass phenomena in the fuel element. The finite element method using asymmetrical 8-node isoparametric elements is used for calculating deformation, stress, and strain in the element and the clad.

Fuel gap conductance is calculated by the Ross and Stoute model [8] taking care of the physical gap existing in the fuel plenum. Plenum gap conductance consists of (a) conduction through solid-solid contact points (b) conduction through solid-gas contact points and, (c) radiation exchange between the element and clad. For plenum gap conductance, the URGAP model of K. Lassmann [9] has been used. The fission gas release is calculated using two methods (a) temperature-dependent release model and (b) physical model based on diffusion and grain growth mechanism.

To estimate the local flux perturbation, the Bessel function is used. The heat transfer from the clad surface to the coolant is calculated using Dittus-Boelter [10] equation. Using the fuel element-clad heat transfer coefficient, new temperature distribution across the fuel and clad is calculated. The corresponding internal gas pressure is calculated using the new temperature distribution and when successive internal gas pressure is within ±5% then the pressure and temperature results are assumed to converge and the iteration is stopped. For improving accuracy, the pellet is divided into 100 rings radially and the fuel temperature and pressure are calculated for each ring. The code is validated against the results of benchmarked codes ELESIM and ELESTRES.

3. Fuel characteristics calculator (FCCAL)

The main thrust of FCCAL [2] is to analyze the plenum gas conductivity with fission gas accumulation and its analytical evaluation. With irradiation of the fuel inside the reactor core, fission noble gases Xenon and Krypton accumulate in the plenum gap which changes the gap conductivity from the initial fuel behavior that is for the Helium-filled during manufacturing. The analytical model is a better approximation over the use of correlations to estimate the effect of noble gases in the plenum gap. The change in conductivity is observable in the fuel in CANDU reactors where on-power refueling is practiced and the old fuel bundles move to higher power generating regions of the core. The fresh fuel along with the old bundles leads to a higher fission rate and hence the release of trapped noble gases towards the plenum. This results in a higher temperature of the old bundles even without an appreciable change in the power. We will discuss these phenomena in the sections below.

3.1 Fission gas release

Fission gases are considered to be released from the fuel when they reach any space that is connected to the free volume within the fuel pin. The released gases accumulate in the fuel-cladding gap, the central void, and porosity within the fuel which communicates directly with the fuel-pin gas space [1]. Cracks or interlinked gas bubbles or pores are an important type of open porosity. The fission gas that has been released from the fuel is responsible for the change in plenum gap conductivity and is assumed to have the following properties. (a) Once the gas is released, the probability of its re-entering the solid from the free volume is negligible (b) the gas pressure in open porosity is equal to that in the free volume of the pin. Because of the insolubility of Xenon and Krypton in solids, there is no effect of plenum fission gas pressure on the rate of gas escape from the fuel (c) while the fission gas contained by the fuel tends to cause swelling, fission gas that has been released promotes shrinkage in the fuel by pressurizing the solid pellets leading to collapse of the internal porosity and bubbles. FCCAL carries out an explicit calculation for changes in gap gas conductivity due to a binary mixture of Helium and Xenon. As the heat generated in the pellet is transferred across the fuel to the coolant, the heat transfer across discontinuities is calculated in the following steps. (i) Heat transfer from the meat of the pellet to the pellet surface. It is estimated by the heat transfer coefficient of the natural uranium oxide pellet (h_P). (ii) Heat transfer across the plenum gap and the Zircalloy clad $\left(h_{Gg} + h_{Gs} + h_s\right)$. h_{Gg} is the heat transfer coefficient due to plenum gas, h_{Gs} is the heat transfer coefficient of the solid-solid contact points between the pellet and the sheath, and h_s is the heat transfer of the Zircalloy sheath. (iii) Heat transfer from the clad outer surface to the coolant is estimated by the heat transfer coefficient of the coolant film near the fuel surface (h_{cf}). Total heat transfer coefficient (h_T) is the sum of terms in (i), (ii), and (iii) i.e.

$$h_T = h_P + h_{Gg} + h_{Gs} + h_s + h_{cf} \tag{1}$$

3.1.1 Fuel pellet conductivity and temperature calculation

PHWR fuel is made of ceramic containing UO_2 with 0.7% U-235. It has poor heat conductivity properties as compared to carbide or metallic uranium. The heat transfer coefficient is dependent upon temperature as well as fission product accumulating inside the pellet. Moreover, as the fuel undergoes irradiation, cracks develop which changes the conductivity. A widely used correlation for the calculation of temperature-dependent pellet conductivity is as follows [11].

$$0 < T \le 1650°C$$

$$h_P = \eta \frac{B_1}{B_2 + T} + B_3 e^{(B_4 T)} \tag{2}$$

$$1650 \le T \le 2940°C$$

$$h_P = \eta \left[B_5 + B_3 e^{B_4 T} \right] \tag{3}$$

$$\eta = \left[\frac{1 - \beta \left(1 - \frac{\rho}{P_{TD}} \right)}{1 - \beta (1 - 0.95)} \right] \tag{4}$$

Where η is the porosity factor and $\beta = 2.58 - 0.58 \times 10^{-3} \times T$. The constants for a different fuel types are shown in **Table 1**.

This correlation is based on the data pooled from ten sources and an analytical expression is generated based on this data. The integral of UO_2 thermal conductivity between $0°C$ and the melting point $2850°C$ is analytically determined in MATPRO. Assuming that the electronic contribution $B_3 e^{(B_4 T)}$ has the value of 2×10^{-3} w/cmKat 1500°C, a least-squares value of 97w/cm is obtained for the integral of h_p from 0°C to the melting point. Data points were fit to an equation including a temperature-dependent, modified Loeb porosity correction.

3.1.2 Equivalent conductivity and temperature drop across plenum gap

Heat transfer coefficient h_{Gg} due to fission gas accumulating in the gap between the pellet and the sheath is a function of fission gas diffusing from the pellet towards the plenum gap. For fresh fuel, the conductivity is a function of helium thermal conductivity but the fission gas changes the gap conductivity. Change in the composition of the plenum gas is a function of fission gas accumulating in the plenum and it is estimated using the industry standard for estimation of the fraction of Xe and Kr [12] diffusing to the plenum as shown in **Table 2**.

Fuel	B_1 w/cm	$B_2°C$	B_3w/cm°C	$B_4°C^{-1}$	B_5w/cm°C
UO_2	40.4	464	1.216×10^{-4}	1.867×10^{-3}	0.0191
$(U, Pu)O_2$	33.0	375	1.540×10^{-4}	1.710×10^{-3}	0.0171

Table 1.
Correlation constants for different fuel types.

Lower temperature limit	Higher temperature limit	Fraction
<1400°C	—	0.05
1400°C	1500°C	0.10
1500°C	1600°C	0.10
1600°C	1700°C	0.10
1700°C	1800°C	0.10
1800°C	2000°C	0.10
>2000°C	—	0.98

Table 2.
Temperature-dependent fraction of fission gas Xe and Kr in the plenum gap.

The burnup-dependent yield of the fission product noble gases is input through a data file in FCCAL. To estimate the cumulative effect of fission gas in the plenum gap n-component gas mixture model is applied [13–15] as shown in Eq. (5).

$$\lambda_{mix} = \sum_{i=1}^{n} \frac{\lambda_t}{1 + \sum_{j=1}^{n} \varphi_{ij} \frac{X_i}{X_j}} \tag{5}$$

Where, λ_{mix} and λ_t are the thermal conductivities of the mixture gas and the individual component gases respectively, X_i and X_j are the mole fractions of the component gases and φ_{ij} is constant. For the binary gas mixture consisting of He-Kr or He-Xe Eq. (1) may be written as:

$$\lambda_{mix} = \frac{\lambda_1}{1 + \varphi_{12} \frac{X_2}{X_1}} + \frac{\lambda_2}{1 + \varphi_{21} \frac{X_1}{X_2}} \tag{6}$$

Where, subscript 1 is for heavier gas of the binary mixture. Values of φ_{ij} [16] is shown in **Table 3**.

The values of φ_{ij} for the component, mixture gases are independent of composition and temperature as shown by Gambhir and Saxena [17, 18] and are given by the following expression.

$$\frac{\varphi_{ij}}{\varphi_{ji}} = \frac{\lambda_i}{\lambda_j} \frac{59M^2 + 88M + 150}{150M^2 + 88M + 59} \tag{7}$$

$$M = \frac{M_2}{M_1} \tag{8}$$

These formulas are important because we can estimate the λ_{mix} of multi-component gas mixtures if the thermal conductivity values of the corresponding binary and pure components are known. Moreover, these formulas help us to obtain λ_{mix} value at high temperature from knowledge of pure λ values at that temperature. Thus φ_{ij} values determined at some lower temperature can be used to calculate λ_{mix} at some higher temperature.

$\varphi_{He-Xe} = 3.4284$	$\varphi_{Xe-He} = 0.3849$
$\varphi_{He-Kr} = 2.7863$	$\varphi_{Kr-He} = 0.4909$

Table 3.
Mixture dependent constants for n (=2) component gas mixture equivalent conductivity formula.

3.1.3 Zircalloy sheath conductivity and temperature drop across the coolant

Zircalloy sheath heat transfer coefficient h_s is calculated based on Eq. (9) [12].

$$h_s = [7.51 + 2.09 \times 10^{-2} - 1.45 \times 10^{-5}T^2 + 7.67 \times 10^{-9}T^3] \qquad (9)$$

Where, h_s is in W/mK and T is in K. Coolant heat transfer coefficient h_{cf} is calculated by the Dittus-Boelter correlation [10] as shown in Eq. (10).

$$Nu = \left(\frac{hDe}{k}\right)_b = C\left(\frac{DeG}{\mu}\right)_b^{0.8} \left(\frac{c_p\mu}{k}\right)_b^n \qquad (10)$$

Where,
h = heat transfer coefficient, Btu/(hrft^3°F);
De = equivalent diameter, ft.;
k = thermal conductivity of fluid, Btu/hr. ft. °F;
c_p = specific heat of fluid, Btu/lb.;
G = mass velocity, lb./hr. ft^2;
μ = fluid viscosity, lb./hr. ft.;
$\frac{De\,G}{\mu} > 10,000$ and $\frac{L}{De} > 60$;
B = bulk conditions.
All other terms have their standard notational convention.

3.1.4 FCCAL code methodology

FCCAL is written in Fortran computer programming language. It is used to calculate fuel centerline temperature, fuel surface temperature, and average coolant temperature in steady-state as well as transient conditions as a function of coolant flow rate (Kg/s), bundle power (Kw), Fuel burnup (MWD/TeU), average channel power (Kw) and actual coolant temperature. For the start of the iteration, a guess value of fuel centerline temperature is assumed taking into account the fact that the fuel centerline temperature cannot be less than the coolant temperature for a reactor operating at steady power. The heat generated in the fuel is taken away by the coolant and transferred to the steam drum and further goes on to generate electricity. Equivalent thermal conductivity across the pellet, plenum gap, clad, and heat transfer to the coolant is calculated. The results are compared with the actual instrument measurements and the guess temperature is accordingly re-evaluated. The guess temperature is accepted to be correct when the error in the code computed values and the instrumented value is within ±2°C. The instrumented temperature is measured from platinum resistance temperature detectors. An error of this magnitude is acceptable as the instrument measurement error is ±2°C.

4. Results

The thermal conductivity of the binary mixture of He-Xe and He-Kr is calculated using the code and the temperature profile is shown in **Figure 1**. The fuel assembly that is analyzed has the same power generation rate and is computed for the same fuel irradiation history.

As the yield of Kr is small as compared to Xe yield hence its effect on the total temperature is small and He dominates the equivalent heat transfer characteristics. For fuel-producing nearly equal power but different irradiation history, we observe that the assembly has a different temperature (**Figure 2**). This is attributable to poor

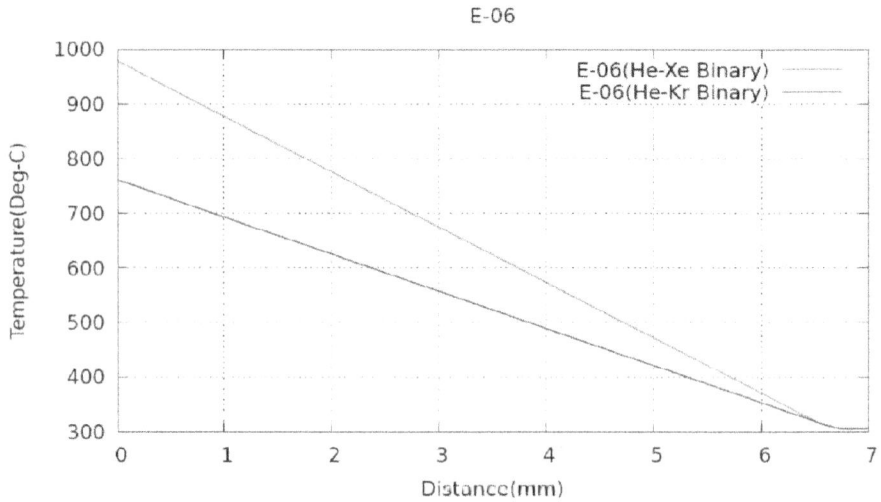

Figure 1.
The temperature profile of fuel assembly for He-Xe binary and He-Kr binary.

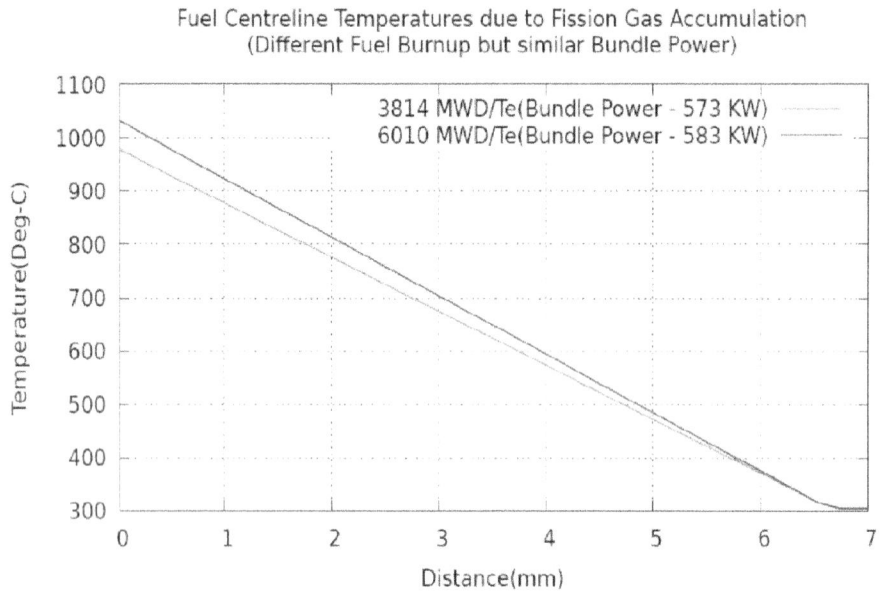

Figure 2.
Assembly producing similar power but different irradiation history and hence different amount of xenon gas buildup in the plenum gap.

heat conduction properties of the He-Xe binary and relative dominance of Xe in the equivalent conductivity. The fuel parameters computed using FCCAL are compared with MATPRO-10 typical parameter values and agree with MATPRO predictions.

5. Conclusion

A model for calculation of fuel temperature profile using binary gas mixture is presented in this chapter along with a discussion of two benchmarked codes for fuel characteristics evaluation. Computing the effect of fission gas products on the overall fuel temperature is presented with the rod irradiation phenomena. From the analysis, it is clear that the code FCCAL can be used for the calculation of fuel centerline temperature, fuel surface temperature, and average coolant temperature of the Pressurized Heavy Water Reactor (PHWR). A better approximation can be obtained by incorporating the fuel cracking and deformation. The effect of fuel heat loss due to irradiation although negligible in steady-state assumes significance in severe transients.

Acknowledgements

The author acknowledges that he is an employee of the atomic power plant and is involved in the PHWR operational aspects.

Author details

Alok Jha[1,2]

1 Homi Bhabha National Institute, Mumbai, India

2 Tarapur Atomic Power Station, Thane, India

*Address all correspondence to: jhaalok1984@gmail.com

IntechOpen

References

[1] Olander, D.R., 1976. Fundamental aspects of nuclear reactor fuel elements. 1 (1976). ERDA Techn. Information Center.

[2] Jha, Alok, et al. "Calculation of fuel temperature profile for heavy water moderated natural uranium oxide fuel using two gas mixture conductance model for noble gas Helium and Xenon." Nuclear Engineering and Technology 52.12 (2020): 2760-2770.

[3] Rest, J. GRASS-SST: a comprehensive, mechanistic model for the prediction of fission-gas behavior in UO 2-base fuels during steady-state and transient conditions. No. NUREG/CR–0202. Argonne National Lab., 1978.

[4] Beyer, C. E., et al. GAPCON-THERMAL-2: a computer program for calculating the thermal behavior of an oxide fuel rod. No. BNWL-1898. Battelle Pacific Northwest Labs., Richland, Wash.(USA), 1975.

[5] Hann, C. R., C. E. Beyer, and L. J. Parchen. GAPCON-THERMAL-1: a computer program for calculating the gap conductance in oxide fuel pins. No. BNWL"1778. Battelle Pacific Northwest Labs., 1973.

[6] Das H, Bhardwaj SA. Fuel Design Analysis Code—FUDA, PPEd Internal Report, IAEA TecDoc 998. 1981. pp. 172-243

[7] Notley, M. J. F. "ELESIM: A computer code for predicting the performance of nuclear fuel elements." Nuclear Technology 445 (1979): 445-450.

[8] Ross, A. M., and R. L. Stoute. Heat transfer coefficient between UO 2 and Zircaloy-2. No. AECL–1552. Atomic Energy of Canada Limited, 1962.

[9] Lassmann K. URANUS—A computer program for the thermal and mechanical analysis of the fuel rods in the nuclear reactor. Nuclear Engineering and Design. 1978;45:325-342

[10] Dittus F, Boelter L. Heat Transfer in Automobile Radiators of the Tubular Type. Berkerly: University of California Publication on Engineering; 1930. p. 371

[11] Reymann, G. A. Matpro–version 10: a handbook of materials properties for use in the analysis of light water reactor fuel rod behavior. No. TREE-NUREG-1180. Idaho National Engineering Lab., Idaho Falls (USA), 1978.

[12] Todreas NE, Kazimi MS. Nuclear Systems: Thermal Hydraulic Fundamentals. USA: Taylor and Francis; 1989

[13] Mason, E. A., and S. C. Saxena. "Approximate formula for the thermal conductivity of gas mixtures." The Physics of fluids **1**.5 (1958): 361-369.

[14] Saxena, S. C., and J. M. Gandhi. "Thermal conductivity of multicomponent mixtures of inert gases." Reviews of Modern Physics **35**.4 (1963): 1022.

[15] Von Ubisch, H., S. Hall, and R. Srivastav. Thermal conductivities of mixtures of fission product gases with helium and with argon. No. A/CONF. 15/P/143. AB Atomenergi, Stockholm, 1959.

[16] von Ubisch, Hans. "The thermal conductivities of mixtures of rare gases at 29°C and at 520°C" Arkiv Fysik 16 (1959).

[17] Saksena, M. P., and S. C. Saxena. "Thermal conductivity of mixtures of monoatomic gases." Proc. Natl. Inst. Sci. India Phys. Sci 26 (1963).

[18] Gambhir RS, Saxena SC. Translational thermal conductivity and viscosity of multicomponent gas mixtures. Trans. Faraday Society. 1964; **60**:38-44

Chapter 9

Fault Detection by Signal Reconstruction in Nuclear Power Plants

Ibrahim Ahmed, Enrico Zio and Gyunyoung Heo

Abstract

In this work, the recently developed auto associative bilateral kernel regression (AABKR) method for on-line condition monitoring of systems, structures, and components (SSCs) during transient process operation of a nuclear power plant (NPP) is improved. The advancement enhances the capability of reconstructing abnormal signals to the values expected in normal conditions during both transient and steady-state process operations. The modification introduced to the method is based on the adoption of two new approaches using dynamic time warping (DTW) for the identification of the time position index (the position of the nearest vector within the historical data vectors to the current on-line query measurement) used by the weighted-distance algorithm that captures temporal dependences in the data. Applications are provided to a steady-state numerical process and a case study concerning sensor signals collected from a reactor coolant system (RCS) during start-up operation of a NPP. The results demonstrate the effectiveness of the proposed method for fault detection during steady-state and transient operations.

Keywords: auto associative kernel regression, auto associative bilateral kernel regression, condition monitoring, dynamic time warping, signal reconstruction, fault detection, nuclear power plant

1. Introduction

In a nuclear power plant (NPP), accurate situation awareness of key systems, structures, and components (SSCs) is important for safety, reliability, and economics which are key drivers for operation. However, faults and failures can occur in sensors and equipment, which can lead to unexpected shutdown of the power reactors. Such situations may compromise the safety and reliability of the SSCs and result in risk and economic losses that may amount to hundreds of thousands of Euros [1]. For example, in United State of America, the economic loss as a result of shutting down a NPP is approximately $1.25 million per day [2]. Thus, if unnecessary shutdown of the system as a result of faults and failures can be prevented, economic loss due to shutdown can be minimized. Therefore, it is of paramount importance to improve the situation awareness of SSCs in NPPs in order to ensure that their faults and failures are detected early, which can be achieved through on-line signal analysis techniques [3–6], and fault detection and diagnosis (FDD) methods [7–13]. There are several techniques of signal analysis, fault detection, and fault diagnostics, which can be classified into two main categories: model-based

[14–17] and data-driven [18–22] methods. The model-based approaches require understanding of the target system's physical structure in order to develop a mathematical model of system response for the purpose of FDD. Data-driven techniques, instead, use the historical data measured by the installed sensors and collected overtime during system' operation to develop an empirical model. In both cases, the developed model can, then, be applied to the target SSC for on-line signal analysis, monitoring, and FDD during operation, from which the condition of the SSC can be retrieved and sent to the human operator/maintenance engineer as alert or alarm, in case of any fault or failure has occurred in the SSC. Based on the status of the SSC, necessary operator action or maintenance intervention can be performed on the SSC to avoid undesired conditions during operation. Adopting these methods in NPP come with several benefits, including [23]:

- Provide system engineers and maintenance staff with necessary information to make informed, cost-effective operations, and maintenance decisions based on the actual condition of the system/equipment.

- Allow early mitigation or corrective actions.

- Reduce the likelihood of unplanned plant trips or power reductions.

- Reduce challenges to safety systems.

- Reduce equipment damage.

- Facilitate the implementation of condition-based predictive maintenance (PdM and CBM) practices.

- Provide significant financial savings, especially if outage duration is reduced.

The recent advancement in data analysis and computational efficiency are motivating the nuclear and other industries to apply CBM for allowing early mitigation, minimizing unplanned shutdown, increasing safety, and reducing maintenance costs. A simple CBM strategy is a scheme that monitors the target component via a fault detection system that continuously collects data from sensors installed on the target component [24], makes a detection decision based on the collected information, provides to operators the condition of the component (normal or abnormal), and triggers an alarm in case of abnormal conditions, which alert the decision makers, for example, stakeholders, operators, and maintenance engineers, for deciding whether or not an intervention on a maintenance action is required on the component.

Figure 1 illustrates an architecture of the fault detection system considered in this work, which is based on an empirical model for signal reconstruction. Typically, as shown in **Figure 1**, a fault detection system is a decision tool based on (i) the model that reconstructs the values of on-line signals expected in normal conditions; and (ii) the residual calculator that analyses the differences between the measured on-line signal values and the reconstructed values, whereby an alarm is triggered if the residuals are statistically deviated from the allowable range representative of normal conditions.

Several empirical models have been developed and used for signal reconstruction. Such techniques include kernel regression (KR)—a special and simple form of Gaussian process regression (GPR) (which has been adapted for signal reconstruction as an auto-associative kernel regression (AAKR) [20] in nuclear industry),

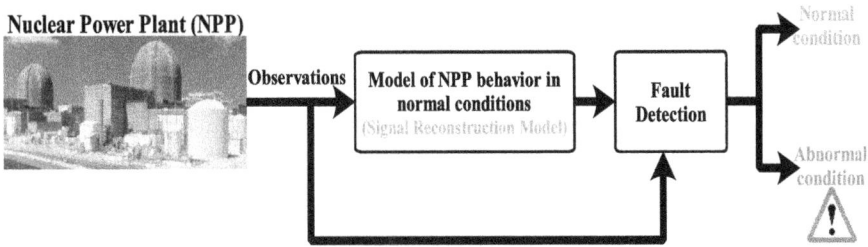

Figure 1.
A typical fault detection system.

auto-associative artificial neural networks (AANNs) [25–27], principal component analysis (PCA) [28–33], multivariate state estimation technique (MSET) [34, 35], Parzen estimation [36], support vector machines [37, 38], evolving clustering method [39], partial least squares (PLS) [40], and fuzzy logic systems [41–45]. However, most data-driven models are developed under steady state plant operation, whereas it is fundamental to have signal validation and monitoring during transient operation as well, considering the fact that most industrial systems' operations are time-varying. Transient operations are any non-steady state, time-varying conditions, which includes start-up, shutdown, and load-following modes of the system, whose time-series data are characterized by an explicit order dependency between observations—a time dimension.

In general, model-based approaches provide valid FDD techniques and are a powerful way to investigate FDD issues in highly dynamic and time-varying systems. However, the high performance of model-based FDD is often achieved at the cost of highly complex process modeling that requires sophisticated system design procedures [46]. Consequently, there is the need for low-complexity data-driven algorithms that could be used for time-varying analysis of the transient operation of the process system. In this respect, AAKR has proven superior to PCA [47] and is less computationally demanding than AANN. AAKR is typically trained to reconstruct the output of its own input under normal conditions. It has been successfully used in actual NPP steady-state operations for instrument channel calibration and condition monitoring [48]. It is a nonparametric technique for estimating a regression function. Unlike parametric models, AAKR relies on the data to determine the model structure.

However, some drawbacks of AAKR, such as spillover effects and robustness issues, can lead to missed alarms or delays in fault detection, and to a difficulty in correctly identifying the sensor variable responsible for a fault that is detected [49]. In order to address these drawbacks, a robust distance measure has been proposed, based on removing the largest elemental difference that contributes to the Euclidean distance metric so as to enable the model to correctly predict sensor values [50]. In [51], a modified AAKR has been proposed, based on a similarity measure between the observational data and historical data, with a pre-processing step that projects both the observed and historical data into a new space defined by a penalty vector.

Although those modifications have improved the AAKR performance, the underlining structure of the AAKR is still based on the traditional unilateral kernel regression and lacks temporal information, which makes its application inappropriate for signal analysis during transient operation because only the current query vector affects the model. Any previous information leading to the current query signals vector is completely ignored. Although this procedure is acceptable and even preferable for many applications, it is not acceptable for transient operations, in which the previous information directly affects the next data point [52, 53].

Recently, a weighted-distance based auto associative bilateral kernel regression (AABKR) for on-line monitoring during process transient operations has been proposed [54] and successfully applied to start-up transient data from an NPP [54, 55]. The AABKR captures both the spatial and temporal information in the data. The time dimension of these kinds of time-series data is, in fact, a structure that provides additional information. The AABKR systematically distributes the weights along the time dimension, using a weighted-distance algorithm that captures temporal dependences in the data [54, 55]. The weighted-distance algorithm uses a derivative-based comparator for the identification of a 'time position index' (the position of the nearest vector, within the historical data vectors, to the current on-line query observation) [54], which directly eliminates the use of on-line time input to the model.

However, when applied to data from steady-state, the performance of the AABKR in terms of correct fault diagnosis (i.e., the identification of the sensor variable responsible for the fault) is not satisfactory, as the fault, in most cases, is detected in both faulty and fault-free sensor signals [54, 55]. After thorough examination, it has been observed that [54]: (1) the AABKR suffers significantly from the spillover effect; (2) the effect is the result of the wrong identification of the 'time position index' by derivatives; and (3) the values of derivatives approximated from a typical steady-state process are, obviously, constants (and nearly zeroes) for most of the data points, particularly, when the process change in time is almost negligible, resulting in wrong identification of the 'time position index'.

It is worth noting that, a correct identification of time position index is crucial for the temporal weighted-distance algorithm that captures the temporal correlation in the data. The consequence of this effect is that, if a fault occurs, it might indeed be detected but, with an incorrect fault diagnosis of the variable responsible for that fault.

Motivated by these observations, we here propose a modified AABKR for efficient on-line monitoring, applicable not only in transient process operations but also in steady-state operations. We develop new algorithms, based on dynamic time warping (DTW), for the identification of temporal dependencies in the data [55]. To evaluate the performance of the proposed methods, we use both synthetic data from a numerical process and real-time data collected from a pressurized water reactor power plant.

2. Problem formulation

We consider a fault detection system designed to monitor the condition of a plant, as shown in **Figure 1,** and the sequence of time-varying observations ordered in time (a time series data). Time is the independent variable, and we assume it to be discrete; thus, time-varying data are a sequence of pairs $[(x_1, t_1); (x_2, t_2); \cdots; (x_M, t_M)]$ with $(t_1 < t_2 < \cdots < t_M)$, where each x_i is a data point in the feature space and t_i is the time at which x_i is observed. The data for more than one signal are sequences of time-varying data points, so long as their sampling rates $(t_i - t_{i-1} = \Delta t = \eta)$ are the same.

With this definition, we assume that:

    i. The sequences of data within an historical time-varying dataset, taken by the sensor in healthy condition, are measured and are available as a memory data matrix, $\mathbf{X} \in \mathbb{R}^{M \times p}$, whose elements, x_{ij}, are functions of the scalar parameter time, t, where $\mathbf{X}$ is a p-dimensional matrix of signals with M observation sequence vectors, and x_{ij} represents the ith observation of the jth signal.

ii. The sequences of data in **X** are large enough to be representative of the plant's normal operating condition.

iii. With r defined as a window length, the sequences of the real-time on-line measurement vectors $\mathbf{X}_q^* \in \mathbb{R}^{r \times p}$ can be collected, with

$$\mathbf{x}_{q_r}^* = \begin{bmatrix} x_{r,1}^* & x_{r,2}^* & \cdots & x_{r,p}^* \end{bmatrix}$$ being the last vector, i.e., the current measurement vector at present time, t.

iv. The sequence of real-time observational matrix, $\mathbf{X}_q^*$, can be updated from current measurement, $\mathbf{x}_{q_r}^*$ backward to the size of the moving window, r, whenever another on-line query vector is available.

On the basis of above descriptions and by using **X**, the objective of the present work is to develop a signal reconstruction model reproducing the plant behavior in normal conditions. Such model receives in input the observed sequence of real-time measurement matrix, $\mathbf{X}_q^*$, whose rth vector is the present measurement, $\mathbf{x}_{q_r}^*$, containing the actual observations of p signals monitored at the present time, t, and produces in output, $\hat{\mathbf{x}}_{q_r}$, the reconstruction values of the signals expected in normal condition. Based on this, the actual plant condition at the present time, t can, then, be determined by the analysis of residuals: a fault is detected if the variations between the observations and the reconstructions are large enough in, at least, one of the signals in comparison to predefined thresholds.

3. Auto associative bilateral kernel regression

3.1 Mathematical framework of AABKR

In AABKR, the validation of the signals at present time, t is based on the analysis of the on-line query pattern, $\mathbf{X}_q^*$, to reconstruct the current vector, $\mathbf{x}_{q_r}^*$, at time, t. The basic idea behind AABKR is to capture both the spatial and temporal correlations in the time-series data (see **Figure 2**), for effective signal reconstruction in the transient operation of industrial systems. The historical memory matrix **X** is reorganized into $\mathcal{A} \in \mathbb{R}^{N \times r \times p}$ sequences of array of N matrices of length r, containing the measurement vectors having $(r-1)$ overlapping between the two

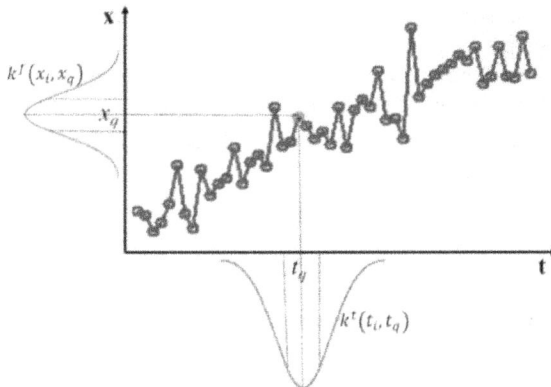

Figure 2.
Graphical representation of the bilateral directions for a time-series.

consecutive time windows, with the r sequence vectors in each matrix of $\mathcal{A}$ array represented as $\mathcal{A}_r^k \in \mathbb{R}^{r \times p}$, where $k = 1, 2, \ldots, N$, and $(N = M - r + 1)$.

The AABKR is expressed in such a way that each neighboring value is weighted on its proximity in space and time. Hence, the mathematical framework of the AABKR is summarized as follows [54, 55]:

1) Feature distance calculation

The feature distance between $\mathbf{x}_{q_r}^*$ and each of the historical memory vectors in matrix $\mathbf{X}$, is computed using the Manhattan distance (L^1-Norm) as

$$d_i\left(\mathbf{X}_i, \mathbf{x}_{q_r}^*\right) = \left\|\mathbf{X}_i - \mathbf{x}_{q_r}^*\right\|_1 = \sum_{j=1}^{p} \left|x_{i,j} - x_{r,j}^*\right| \tag{1}$$

and produces the distance vector, $\mathbf{d} \in \mathbb{R}^{M \times 1}$.

2) Feature kernel quantification

The feature distances calculated above are used to determine the feature weights by evaluating the Gaussian kernel

$$k_i^f = \exp\left(\frac{-d_i^2}{2h_f^2}\right) \tag{2}$$

where h_f is a kernel bandwidth for feature preservation, which controls how much the nearby memory feature vectors are weighted. This leads to the $\mathbf{k}^f \in \mathbb{R}^{M \times 1}$ vector. The superscript/subscript f indicates the feature components.

3) Time position index identification

Here, the time position index, that is, the temporal location of the nearest vector, within the memory vector, to the query vector observation, is determined using a derivative-based comparator. This provides the input to the weighted-distance algorithm in *Step 4*. Instead of directly using the derivative in the prediction to capture the temporal correlation of the data, which might not be a good choice because of process measurement noise, the derivatives are used as a *comparator* to determine the time position index within the memory vectors to which the query data vector is nearest. The derivative-based comparator is described as follows.

The backward-difference first-order derivative approximation of the current historical measurement vector in each matrix $\mathcal{A}_r^k$ based on r data points accuracy with respect to t is the element-by-element derivative:

$$\frac{\partial \mathcal{A}_r^k}{\partial t} = \left[\begin{array}{cccc} \frac{\partial x_{r,1}^k}{\partial t} & \frac{\partial x_{r,2}^k}{\partial t} & \ldots & \frac{\partial x_{r,p}^k}{\partial t} \end{array}\right] \tag{3}$$

whereas, that of the current query vector, $\mathbf{x}_{q_r}^*$ in matrix $\mathbf{X}_q^*$ is the element-by-element derivative:

$$\frac{\partial \mathbf{x}_{q_r}^*}{\partial t} = \left[\begin{array}{cccc} \frac{\partial x_{r,1}^*}{\partial t} & \frac{\partial x_{r,2}^*}{\partial t} & \ldots & \frac{\partial x_{r,p}^*}{\partial t} \end{array}\right]. \tag{4}$$

The first-order derivatives in Eqs. (3) and (4) have been approximated from the data using finite-difference derivative approximation. The backward finite difference derivative approximation is chosen to implement real-time on-line monitoring. The model needs r successive data points to evaluate the derivative of the current data point from the current measurement backward to the size of the moving window r at every sampling time, using backward finite-difference derivative approximation.

From Eqs. (3) and (4), the distance between the derivative of a query vector, $\mathbf{x}_{q_r}^*$ and each kth derivative vector of A_r^k can be calculated by Eq. (5) using the Manhattan distance (L^1 norm):

$$\Delta_k \left(\frac{\partial A_r^k}{\partial t}, \frac{\partial \mathbf{x}_{q_r}^*}{\partial t} \right) = \left\| \frac{\partial A_r^k}{\partial t} - \frac{\partial \mathbf{x}_{q_r}^*}{\partial t} \right\|_1 = \sum_{j=1}^{p} \left| \frac{\partial x_{r,j}^k}{\partial t} - \frac{\partial x_{r,j}^*}{\partial t} \right| \tag{5}$$

This gives the derivative distance vector, $\Delta \in \mathbb{R}^{N \times 1}$.

Then, using the minimum value in Δ, the index $i = \varepsilon$, which indicates the location of the vector in the memory data, $\mathbf{X}$, to which the current query vector $\mathbf{x}_{q_r}$ is closest, can be obtained. The time position index is, therefore, the index at which the Manhattan distance between the derivative of the current query vector and those of the rth vectors in each of the A_r^k is minimized plus the overlapping length between the two consecutive time windows, which is determined as:

$$\varepsilon = \left(\arg \min_{k \in [1:N]} (\Delta_k) \right) + r - 1 \tag{6}$$

4) Temporal weighted-distance algorithm

The temporal weighted-distance algorithm captures the temporal correlations in the data. It calculates the measures that capture the temporal variations in the data. The distance, δ, accounts for the time at which the query vector is observed. This algorithm calculates the temporal correlation of a query input with the memory data, without using the query time input (t_q), and eliminates the direct use of t_q, which becomes indefinite when applied to on-line monitoring. In this way, the effect of the query time input is confined within the time duration of the historical memory data. The distance is calculated based on the assumption that the time-varying historical data collected in building the model were sampled at a constant time interval, η.

Based on the time position index determined in *Step 3*, the temporal weighted-distance algorithm that captures the temporal correlation is formulated as

$$\delta_i = \begin{cases} \delta_\varepsilon, & i = \varepsilon \\ \delta_\varepsilon + (i - \varepsilon).\eta, & i > \varepsilon \& \varepsilon \neq M; \quad i \in [1, M] \\ \delta_\varepsilon + (\varepsilon - i).\eta, & i < \varepsilon \& \varepsilon \neq 1 \end{cases} \tag{7}$$

giving the weighted-distance vector $\delta \in \mathbb{R}^{M \times 1}$:

$$\delta = \begin{bmatrix} \delta_1 & \cdots & \delta_{\varepsilon-2} & \delta_{\varepsilon-1} & \delta_\varepsilon & \delta_{\varepsilon+1} & \delta_{\varepsilon+2} & \cdots & \delta_M \end{bmatrix}^T \tag{8}$$

Once the values of δ_ε and η are known, the other values in Eq. (8) can be determined progressively using Eq. (7). The second and third equations in Eq. (7)

follow arithmetic progression (AP): the first term and the common difference of the two progressions are δ_e and η, respectively. A zero value for the first term of the two progressions, $\delta_e = 0$, has been recommended [55] because the distance of the nearest vector in the memory data to the query vector is close to zero. Whereas the value of the common difference can be arbitrarily selected or taken to be the time interval, η. The other distance values to the right and left of δ_e in Eq. (8) can be progressively calculated using the second and third equations in Eq. (7), respectively. See Appendix C of [55] for the proof of this algorithm (Eq. (7)).

5) Temporal kernel quantification

Having determined the weighted-distance, the kernel weight can be calculated using the Gaussian kernel function:

$$k_i^t = exp\left(\frac{-\delta_i^2}{2h_t^2}\right) \tag{9}$$

where the k_i^t is the ith kernel weight calculated from the temporal weighted-distance, δ_i; the superscript/subscript t indicates the temporal components; h_t is the bandwidth for the time-domain preservation, which can also serve as noise rejection and controls how much the nearby times in the memory vectors are weighted. This gives the vector $\mathbf{k}^t \in \mathbb{R}^{(M \times 1)}$.

6) Adaptive bilateral kernel evaluation

Depending on the magnitude of a fault that occurs in a process, the result of the direct multiplication of the two kernels at $i = e$ could be zero, which would result in an inaccurate model prediction because the model prediction tends to follow the fault occurrence, so the fault would not be detected. To resolve such issue and achieve robust model signal reconstruction, and to reduce the impact of spillover onto other signals when one or more signals is in fault condition, Eq. (10) is formulated [54] adaptively for the combined kernels of Eqs. (2) and (9) as:

$$k_i^{ab} = \begin{cases} k_i^f * k_i^t, & 1 \le i \le M \& i \ne e \\ \dfrac{\left(k_i^f + k_i^t\right)}{2}, & i = e \end{cases} \quad ; \quad i\epsilon[1, M] \tag{10}$$

resulting in the adaptive bilateral kernel vector $\mathbf{k}^b \in \mathbb{R}^{M \times 1}$.

This reduces the effect of the dominance of one feature distance value over another when a fault occurs. The adaptive nature of Eq. (10) is to dynamically compensate for faulty sensor inputs to the bilateral kernel evaluation and always ensure that a larger weight is assigned to the closest vector within the memory data to the query vector, so as to guarantee an approximate signal reconstruction. This reduces the effect of the degeneration of the feature kernel when a fault of high magnitude has occurred.

7) Output estimation

Finally, the adaptive bilateral kernel weights are combined with the memory data vectors to give the predictions as:

$$\hat{x}_{r,j}^{*} = \frac{\sum_{i=r}^{M} k_i^{ab} \cdot x_{i,j}}{\sum_{i=r}^{M} k_i^{ab}} \tag{11}$$

If a normalized adaptive bilateral kernel vector, $\mathbf{w} \in \mathbb{R}^{(M-r+1) \times 1}$ is defined:

$$w_i = \frac{k_i^{ab}}{\sum_{i=r}^{M} k_i^{ab}}, \tag{12}$$

then, Eq. (11) can be rewritten in matrix form to predict all the signals of the query vector simultaneously as:

$$\hat{x}_{q_r}^{*} = \mathbf{w}^T \mathbf{X} \tag{13}$$

where $\mathbf{X} \in \mathbb{R}^{(M-r+1) \times p}$.

8) Fault detection

After training of the model, the root mean square error (RMSE) on the predictions of the fault-free validation dataset can be calculated using residuals $\left(e_{q_r} = \mathbf{x}_{q_r} - \hat{\mathbf{x}}_{q_r}^{*} \right)$ between the actual values and the predicted values of the validation dataset, and can be used to set the threshold limit for fault detection in each signal as follows:

$$T_j^D = 3 * RMSE_j. \tag{14}$$

Because the residuals can be assumed to be Gaussian and randomly distributed with a mean of zero and variance of $RMSE_j{}^2$, a constant value equal to 3 has been selected in [54] to minimize the false alarm rate and ensure that a fault is detected when the residuals exceed the threshold.

3.2 Analysis of the limitation of the AABKR

In this section, we analyze the limitation of the AABKR described in Section 3.1, in terms of signal reconstruction from faulty sensor signals. The major limitation can be understood from the description presented as follows.

We observed that, in an extreme, limit or worst case scenario, where the fault deviation intensity in a faulty sensor signal is significant, the feature distance vector degenerates and tends to zero (i.e., $\mathbf{k}^f \approx 0$), so that the signal reconstructed by the AABKR model is bound to be:

$$\hat{x}_{q,j}^{*} = x_{\ell,j} \tag{15}$$

This observation can be understood better by the following analysis.
Recall the reconstructed output from a weighted average of Eq. (11), re-written as:

$$\hat{x}_{q,j}^{*} = \frac{\sum_{i=1}^{M} \left(k_i^f \circledast k_i^t \right) x_{i,j}}{\sum_{i=1}^{M} \left(k_i^f \circledast k_i^t \right)} \tag{16}$$

where, $k_i^f \circledast k_i^t = k_i^{ab}$ is the adaptive bilateral kernel evaluated at $\mathbf{x}_i$. The symbol $\circledast$ represents the bilateral kernel combination operator that combines the feature and temporal kernels, given by Eq. (10).

Applying the properties of limit to Eq. (16), we have:

$$\lim_{k^f \to 0} \left(\hat{x}_{qj}^* \right) = \lim_{k^f \to 0} \left(\frac{\sum_{i=1}^{M} \left(k_i^f \circledast k_i^t \right) x_{i,j}}{\sum_{i=1}^{M} \left(k_i^f \circledast k_i^t \right)} \right) \tag{17}$$

Equation (17) can be re-written as:

$$\lim_{k^f \to 0} \left(\hat{x}_{qj}^* \right) = \frac{\lim_{k^f \to 0} \left(\sum_{i=1}^{M} \left(k_i^f \circledast k_i^t \right) x_{i,j} \right)}{\lim_{k^f \to 0} \left(\sum_{i=1}^{M} \left(k_i^f \circledast k_i^t \right) \right)} \tag{18}$$

provided that:

$$\lim_{k^f \to 0} \left(\sum_{i=1}^{M} \left(k_i^f \circledast k_i^t \right) \right) \neq 0 \tag{19}$$

Note that, judging from the Eq. (10), Eqs. (18) and (19) hold valid. Hence, the limit in Eq. (18) can be simplified as:

$$\lim_{k^f \to 0} \left(\hat{x}_{qj}^* \right) = \frac{\lim_{k_1^f \to 0} \left(\left(k_1^f \circledast k_1^t \right) x_{1,j} \right) + \lim_{k_2^f \to 0} \left(\left(k_2^f \circledast k_2^t \right) x_{2,j} \right) + \cdots + \lim_{k_M^f \to 0} \left(\left(k_M^f \circledast k_M^t \right) x_{M,j} \right)}{\lim_{k_1^f \to 0} \left(k_1^f \circledast k_1^t \right) + \lim_{k_2^f \to 0} \left(k_2^f \circledast k_2^t \right) + \cdots + \lim_{k_M^f \to 0} \left(k_M^f \circledast k_M^t \right)} \tag{20}$$

But, from the adaptive bilateral combination of Eq. (10):

$$\lim_{k_i^f \to 0; i \neq \varepsilon \& 1 \leq i \leq M;} \left(k_i^f \circledast k_i^t \right) = 0 \tag{21}$$

and

$$\lim_{k_i^f \to 0, i = \varepsilon} \left(k_i^f \circledast k_i^t \right) = 0.5 \tag{22}$$

Hence Eq. (20) becomes

$$\lim_{k^f \to 0} \left(\hat{x}_{qj}^* \right) = \frac{\lim_{k_i^f \to 0, i = \varepsilon} \left(\left(k_i^f \circledast k_i^t \right) x_{i,j} \right)}{\lim_{k_i^f \to 0, i = \varepsilon} \left(k_i^f \circledast k_i^t \right)} = x_{\varepsilon,j} \left(\frac{0.5}{0.5} \right) \tag{23}$$

Thus,

$$\lim_{k^f \to 0} \left(\hat{x}_{qj}^* \right) = x_{\varepsilon,j} \tag{24}$$

This implies that, in a limit case of faulty sensor query signal, x_{qj}^*, the reconstructed query signal, $\hat{x}_{qj}^*$ is equal to the historical (memory) data point, $x_{\varepsilon,j}$ located at the identified time position index, ε.

From the above analysis, it is clear that the fault detection capability of the AABKR largely depends on the accuracy of the time position index identification algorithm. If the time position index has been identified correctly, the fault can be detected even though the fault deviation intensity is large and tends to infinity. However, if the time position index has been identified wrongly, the fault might or might not be detected depending on the identified index and the intensity of the fault. The consequence of this situation, even if the fault has been detected, is that the sensor signal responsible for the fault might not be diagnosed correctly which might lead to wrong diagnosis of the system under consideration. In this regards, a more robust approach for time position index identification is proposed and discussed in the next section.

4. Modified AABKR

In this section, the modified AABKR is presented. The framework of the proposed method is depicted in **Figure 3**. It comprises the steps of calculating the feature distance that captures the spatial variation in the data; calculating the feature kernel weights based on the calculated feature distance; identifying the time position index using DTW technique; computing the temporal weighted-distance that captures the temporal variation and dependencies in the data, based on the time position index; calculating the temporal kernel weights, based on the calculated weighted-distance; and evaluating the adaptive bilateral kernel that computes the combined kernels and dynamically compensates for faulty sensor inputs to the bilateral kernel evaluation, and then, makes the prediction using a weighted average for the purpose of fault detection. Only the modifications to the original AABKR are discussed in this section. The basics of the DTW technique is first presented in Section 4.1. Then, the developed methods based on the DTW are discussed in Section 4.2. Finally, a demonstration of the developed identification methods is showcased in Section 4.3.

4.1 Dynamic time warping

Dynamic time warping (DTW) is a technique for finding an optimal alignment between two time-dependent sequences. This technique uses a dynamic programming approach to align the time-series data [56]. Suppose we have two time-series sequences of values taken from the feature space, $\mathbf{x} = [\,x_1 \quad x_2 \quad \cdots \quad x_L\,]$ and $\mathbf{y} = [\,y_1 \quad y_2 \quad \cdots \quad y_M\,]$, of length L and M, respectively. To align these sequences using DTW, an $L \times M$ matrix, $\mathbf{D}$, is first established, where the element $d_{l,m}$ of $\mathbf{D}$ is a local distance measured between the points x_l and x_m, usually called *cost function* since the DTW technique is based on the dynamic programming algorithm. Thus, the task of optimal alignment of these sequences is the arrangement of all sequence

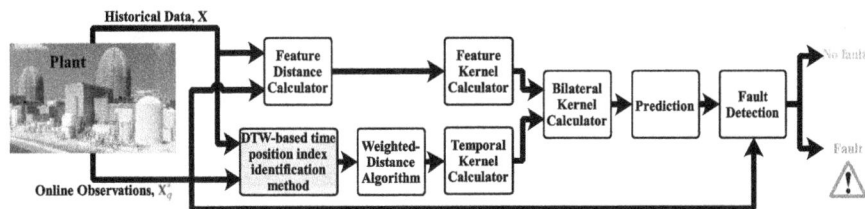

Figure 3.
Framework of the modified AABKR for fault detection system.

points for minimizing the *cost function*. Once the local cost matrix has been built, the algorithm finds the alignment path which runs through the low-cost area—valley on the matrix. The alignment path, usually called the warping path, is a sequence of points, $\mathbf{w} = [\, w_1 \quad w_2 \quad \cdots \quad w_P \,]$ and defines the correspondence of an element x_l to y_m with $w_p = (l_p, m_p) \in [1 : L] \times [1 : M]$ for $p \in [1 : P]$ satisfying the following three criteria [57, 58]:

1. Boundary condition: $w_1 = (1,1)$ and $w_P = (L, M)$.

2. Monotonicity condition: $l_1 \le l_2 \le \cdots \le l_P$ and $m_1 \le m_2 \le \cdots \le m_P$.

3. Step size condition: $w_{p+1} - w_p \in \{(1, 0), (0, 1), (1, 1)\}$ for $p \in [1 : L - 1]$.

The total cost of a warping path $\mathbf{w}$ between $\mathbf{x}$ and $\mathbf{y}$ with respect to the local cost matrix (which represents all pairwise distances) is:

$$d_{\mathbf{w}}(\mathbf{x}, \mathbf{y}) = \sum_{p=1}^{P} d(w_p) = \sum_{p=1}^{P} d\left(x_{l_p}, y_{m_p}\right) \tag{25}$$

Moreover, an optimal warping path between $\mathbf{x}$ and $\mathbf{y}$ is a warping path, $\mathbf{w}^*$, that has minimal total cost among all possible warping paths. The DTW distance, DTW $(\mathbf{x}, \mathbf{y})$ between $\mathbf{x}$ and $\mathbf{y}$ is, then, defined as the total cost of $\mathbf{w}^*$:

$$\mathrm{DTW}(\mathbf{x}, \mathbf{y}) = d_{\mathbf{w}^*}(\mathbf{x}, \mathbf{y}) = min\left\{d_{\mathbf{w}}(\mathbf{x}, \mathbf{y}), w \in \mathbf{W}^{L \times M}\right\} \tag{26}$$

where $\mathbf{W}^{L \times M}$ is the set of all possible warping paths.

To determine the optimal warping path, one could test every possible warping path between $\mathbf{x}$ and $\mathbf{y}$, which would however very be computationally intensive. Therefore, the optimal warping path can be found using dynamic programming by building an accumulated cost matrix called global cost matrix, $\mathbf{G}$, which is defined by the following recursion [57]:

First row:

$$\mathbf{G}(1, m) = \sum_{k=1}^{m} d(x_1, y_k), \quad m \in [1, M] \tag{27}$$

First column:

$$\mathbf{G}(l, 1) = \sum_{k=1}^{m} d(x_k, y_1), \quad l \in [1, L] \tag{28}$$

All other elements:

$$\mathbf{G}(l, m) = d(x_l, y_m) + min \left\{ \begin{array}{l} \mathbf{G}(l - 1, m - 1), \\ \mathbf{G}(l - 1, m), \\ \mathbf{G}(l, m - 1) \end{array} \right\}, l \in [1, L] \,\&\, m \in [1, M] \tag{29}$$

This means that, the accumulated global distance is the sum of the distance between the current elements and the minimum of the accumulated global distances of the neighboring elements. The required time for building matrix $\mathbf{G}$ is $O(LM)$.

Having determined the accumulated cost matrix $\mathbf{G}$, obviously, the DTW distance between $\mathbf{x}$ and $\mathbf{y}$ is simply given by Eq. (30):

$$DTW(\mathbf{x}, \mathbf{y}) = \mathbf{G}(L, M).\qquad(30)$$

Once the accumulated cost matrix has been built, the optimal warping path could be found by backtracking from the point $w_P = (L, M)$ to the $w_1 = (1, 1)$, using a greedy strategy.

4.2 DTW-based time position index identification approaches

One of the significant steps that requires modification in AABR is the identification of the time position index. The goal here is that the identification of the time position index should be solely based on the feature data and must be freed from the use of the derivatives, in order to improve the monitoring performance of the AABKR during steady-state operations. To achieve this goal, we developed two approaches based on the DTW algorithm described in Section 4.1 for the identification of the time position index. The two approaches are described in the following subsections.

4.2.1 First approach: based on the generated subsequences of the memory data

For simplicity of description, we assume $p = 1$ (single signal) and present the description of the algorithm by referring to **Figure 4**, where $\mathbf{X} \in \mathbb{R}^{M \times p}$ and $\mathbf{X}_q^* \in \mathbb{R}^{r \times p}$ are memory data and query data, respectively. We also assume that the memory data, $\mathbf{X}$ has already been reorganized into an array containing N matrices each of length r with $(r - 1)$ overlapping between them, as described in Section 3.1 where $(N = M - r + 1)$. By using DTW, the goal is to find a subsequence $A_r^k[a^* : \varepsilon] \in \mathbb{R}^{r \times p} = [x_{a^*}, x_{a^*+1}, \cdots x_\varepsilon]$, with $1 \leq a^* \leq \varepsilon \leq M$, that minimizes the DTW distance to $\mathbf{X}_q^*$ over all N matrices generated from $\mathbf{X}$. Note that, A_r^k is a kth subsequence in the generated array, where $k = 1, 2, \ldots, N$. Thus, the DTW distance between $\mathbf{X}_q^*$ and each of the kth matrix, A_r^k in $\mathcal{A}$ can be determined by:

$$DTW\left(\mathbf{X}_q^*, A_r^k\right) = \mathbf{G}(r, r)\qquad(31)$$

which can be calculated from the local cost function matrix, $\mathbf{D}$, using Eqs. (27)–(29) of the DTW algorithm described in Section 4.1. Each element of $\mathbf{D}$

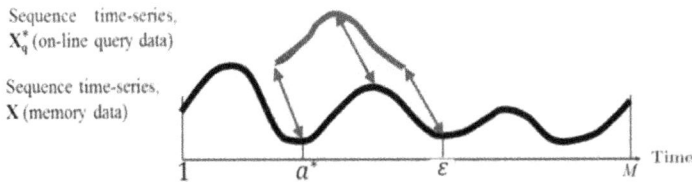

Figure 4.
Alignment between two time-dependent data: Sequences X_q and X.

is a local distance measured between the points in $\mathbf{X}_q^*$ and $\mathcal{A}_r^k$, which can be calculated by:

$$d[l, m] = d\left(\mathbf{x}_l^*, \mathbf{x}_m\right) = \left\|\mathbf{x}_l^* - \mathbf{x}_m\right\|_1 \tag{32}$$

where $l = 1, 2, \ldots, r$, is the row index of the query data, $\mathbf{X}_q^*$, and $m = 1, 2, \ldots, r$, is the row index of the kth subsequence, $\mathcal{A}_r^k$, of the memory data.

Then, the time position index can be determined as:

$$\varepsilon = \left(\arg\min_{k \in [1:N]} \left\{\mathrm{DTW}\left(\mathbf{X}_q^*, \mathcal{A}_r^k\right)\right\}\right) + r - 1. \tag{33}$$

This index can, then, be used in the weighted-distance algorithm for the calculation of the temporal weighted-distance. This approach is summarized in Algorithm A.2.1 of Appendix A.2.

4.2.2 Second approach: based on the entire memory data

In this approach, instead of calculating the DTW distances between the query input data and each of the subsequence data generated from the memory data, the mapping between the query input data and the memory data can be determined directly, from which the time position index can be obtained. Thus, the generation of the array from the memory data is not required in this case. This approach is described as follows.

First, calculate the local cost function matrix, $\mathbf{D}$ between the query data, $\mathbf{X}_q^*$, and the memory data, $\mathbf{X}$. Having calculated $\mathbf{D}$, the calculation of $\mathbf{G}$ from $\mathbf{D}$ using dynamic programming [57] is a bit modified through the following recursion:

First row:

$$\mathbf{G}(1, i) = d\left(\mathbf{x}_1^*, \mathbf{x}_i\right), \quad i \in [1, M] \tag{34}$$

First column:

$$\mathbf{G}(l, 1) = \sum_{s=1}^{i} d\left(\mathbf{x}_s^*, \mathbf{x}_1\right), \quad l \in [1, r] \tag{35}$$

All other elements:

$$\mathbf{G}(l, i) = d\left(\mathbf{x}_l^*, \mathbf{x}_i\right) + min \begin{cases} \mathbf{G}(l-1, i-1), \\ \mathbf{G}(l-1, i), \\ \mathbf{G}(l, i-1) \end{cases}, l \in [1, r] \& i \in [1, M] \tag{36}$$

leading to an accumulated matrix, $\mathbf{G} \in \mathbb{R}^{r \times M}$.

Finally, the time position index is obtained as follows, using the last row of $\mathbf{G}$:

$$\varepsilon = \arg\min_{i \in [r:M]} \left(\mathbf{G}[r, i]\right) \tag{37}$$

Note that the calculation of $\mathbf{G}$ is the same as earlier discussed, except that, its first row is taken equal to the first row of $\mathbf{D}$ without accumulating [57], as shown in Eq. (34). This is because, our goal is to determine the time position index using the last row of **G** and to minimize the impact of the fault (if it occurs) during the determination of the index. The obtained time position index in Eq. (37) can, then, be used in the weighted-distance algorithm for the calculation of the temporal weighted distance. This approach is summarized in Algorithm A.2.2 of Appendix A.2.

4.3 Demonstration case

To demonstrate the two approaches of time position index identification, we consider a typical univariate time-dependent process:

$$x(t) = -\frac{1}{200,000}t^2 + 8 + g(t). \tag{38}$$

where $g(t)$ is assumed to be an independent and normally distributed Gaussian noise at present time, t with mean equal to zero and standard deviation, 0.08. For simplicity of demonstration, a memory dataset, **X**, consisting of 50 samples (with $t = 1\text{--}50$ s at constant time intervals of $\eta = 1s$), has been generated from the above process. By setting the window size, $r = 10$, a query input of length r is, then, generated. The actual location of the query input within the memory data is from $t = 36\text{--}45$ s, where the time position index of the current data point is at time $t = 45$ s. The goal is to automatically locate this index using the proposed methods. The generated memory data is plotted as shown in **Figure 5**, where the location of the query input is indicated in blue color.

With respect to the first approach, the global matrices, **Gs**, between the query observations and each of the subsequences of the memory data, are first computed and visualized in **Figure 6** for the case of fault-free data. Next, a fault is added to the data point at present time t ($t = 45$ s) within the query data pattern and, then, the global matrices are recomputed as depicted in **Figure 7**. The DTW distances from global matrices are presented in **Figure 8** for both cases of fault and no fault. It can be seen that, the index at which the present data point is closest to has been identified in both fault-free and faulty cases. From **Figure 8** and by using Eq. (33), the time position index is $\varepsilon = 36 + (10 - 1) = 45$.

With respect to the second approach, the global matrix between the query observation and memory data is first computed and visualized in **Figure 9** for the fault-free case. Next, a fault is added to the data point at present time t ($t = 45$ s) within the query data pattern and, then, the global matrix is recomputed as shown in **Figure 10**. Finally, the last row of the global matrix is used to determine the index

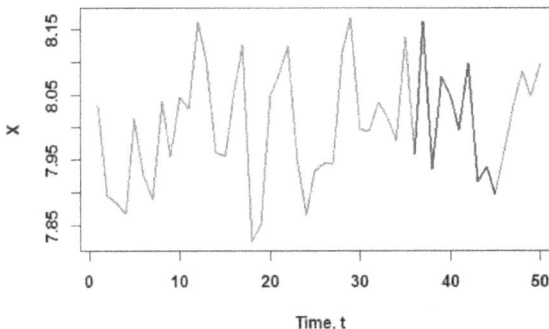

Figure 5.
Memory data (in red) and query input (in blue, located at t = 36–45 s).

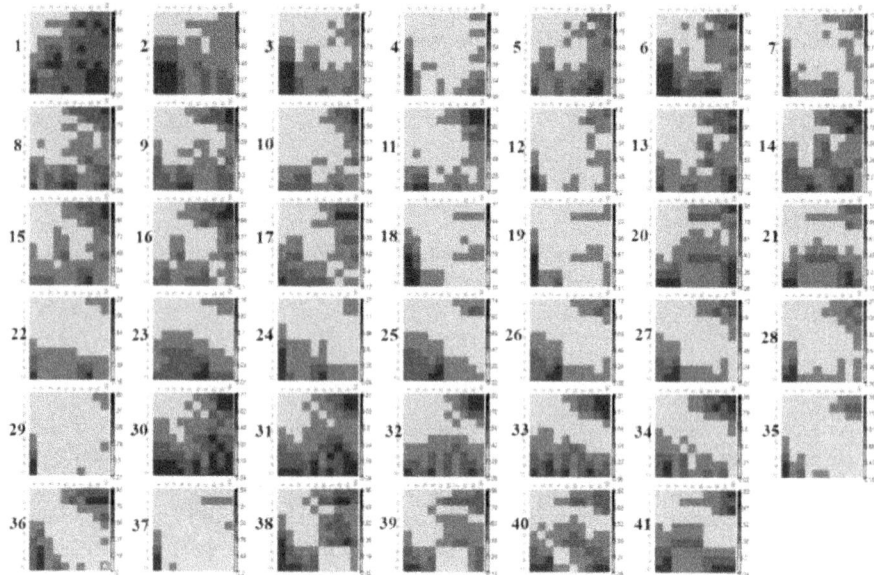

Figure 6.
Global matrices, G, between X^*_q and subsequences of X (fault-free case).

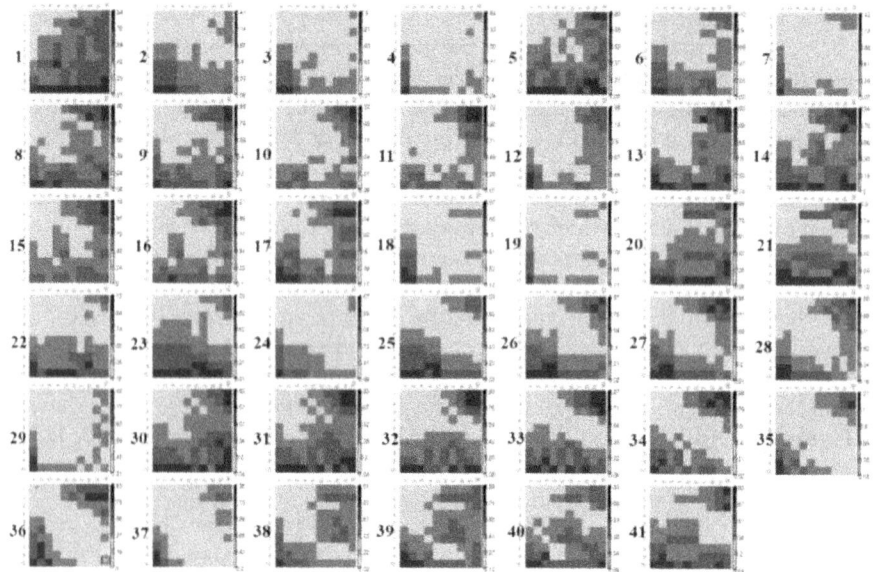

Figure 7.
Global matrices, G, between X^*_q and subsequences of X (fault case).

in both cases of fault and no fault, using Eq. (37). The locations identified in both cases ($\varepsilon = 45$) are marked in red square box as shown in **Figures 9** and **10**.

We observe that if the fault deviation intensity increases, the identification accuracy of the second approach decreases and the time position index would not be identified correctly, whereas, the first approach would still identify the time position index correctly but with high computational demand. That is, the second approach is less computationally demanding than the first approach but it is less accurate.

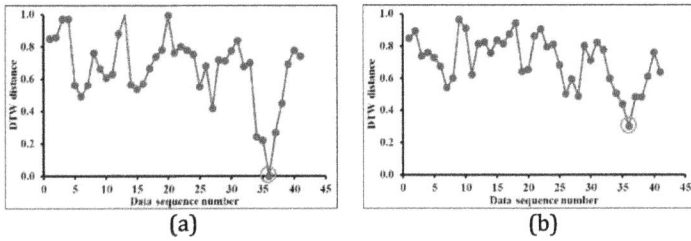

Figure 8.
DTW distances between X^*_q and subsequences of X. (a) Fault-free case and (b) faulty case.

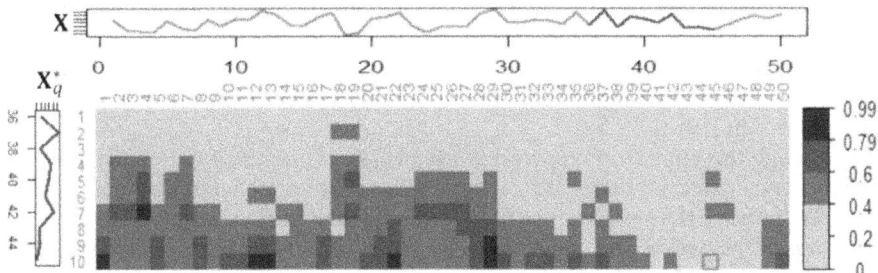

Figure 9.
Global matrix, G, between X^*_q and memory data, X (fault-free case).

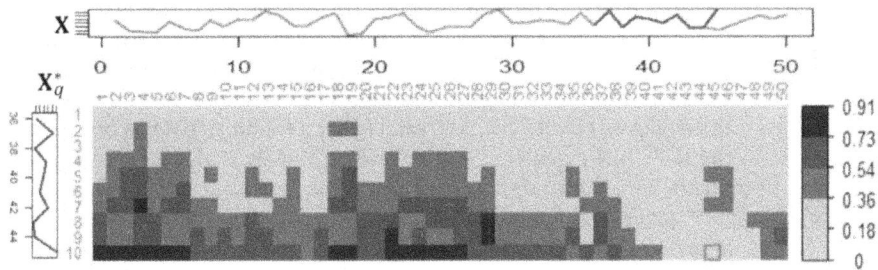

Figure 10.
Global matrix, G, between X^*_q and memory data, X (fault case).

5. Applications

In this section, a typical steady-state numerical example taken from a literature is first used to evaluate the fault detection capability of the proposed signal reconstruction methods in steady-state operation and, then, applied to the transient start-up operation of a nuclear power plant.

5.1 Validation on the steady-state process

A typical steady-state numerical example [54], mimicking a typical industrial system, is considered to evaluate the performance of the proposed method in steady-state operation. The model of the process is:

$$
\begin{bmatrix} x_1 \\ x_2 \\ x_3 \\ x_4 \\ x_5 \\ x_6 \end{bmatrix} = \begin{bmatrix} -0.2310 & -0.0816 & -0.2662 \\ -0.3241 & 0.7055 & -0.2158 \\ -0.2170 & -0.3056 & -0.5207 \\ -0.4089 & -0.3442 & -0.4501 \\ -0.6408 & 0.3102 & 0.2372 \\ -0.4655 & -0.4330 & 0.5938 \end{bmatrix} \begin{bmatrix} t_1 \\ t_2 \\ t_3 \end{bmatrix} + noise \tag{39}
$$

where t_1, t_2, and t_3 are zero-mean random variables with standard deviations of 1, 0.8, and 0.6, respectively. The noise included in the process is zero-mean with a standard deviation of 0.2, and is normally distributed. To build the model, 1000 samples are generated using such process. The number of simulated faults is 2000, with the data samples generated according to the model above and the fault magnitude being a random number uniformly distributed between 0 and 5. The signal variable in fault is also random uniformly sampled among the six possible variables, as simulated in [59].

Although this application represents a typical steady-state operation, we assume a set of sequential time-series data, assigning a constant time interval of $\eta = 1s$ with $r = 3$. To measure the performance of the propose methods, we employ different measures of the performance metrics (e.g., missed alarm rate (MAR), missed and false alarms rate (M&FAR), true and false alarms rate (T&FAR), true alarm rate (TAR), and fault detection rate (FDR)) as proposed in [54] and briefly defined in Appendix A.3. The purpose of this application is to verify the performance of the proposed method in monitoring during steady-state operation and to compare the results with those of AAKR (see Appendix A.1) and AABKR.

Table 1 and **Figure 11** show the alarm rates of AAKR, AABKR, and the modified AABKR computed from the prediction of the simulated faults. It is interesting to note that although the MAR of AAKR is a bit higher than that of AABKR, the performance of the two models does not differ significantly and both models have suffered from the spillover effects (i.e., the effect that a faulty signal has on the predictions of the fault-free signals) as evident from the values of T&FAR (i.e., the detection of faults in both faulty signal and at least one fault-free signal). Conversely, the performance of the modified AABKR is better than those of the other two methods in terms of TAR (i.e., the detection of fault only in a signal that actually has the fault, without false alarm in other fault-free signals) and T&FAR; hence, the modified AABKR is more resistant to spillover and more robust than both AAKR and AABKR. It can be observed that even though the TAR value of the modified AABKR is larger than those of the other two models, FDR values of the three methods did not differ significantly because of the larger values of T&FAR for AAKR and AABKR (53.4 and 61.8%, respectively). Therefore, it is important to further examine the rate of correct fault diagnosis of the three methods using absolute residual values of the faults successfully detected, which produced the FDR values. **Figure 12** shows the rate of correct fault diagnosis of the three methods. We

Model	MAR	M&FAR	T&FAR	TAR	FDR
AAKR	19.20	1.60	53.40	25.80	80.80
AABKR	5.65	8.26	61.82	24.27	94.35
Modified AABKR	12.46	6.81	28.18	52.55	87.54

Table 1.
Alarm rates (%) of validation on the numerical steady-state process.

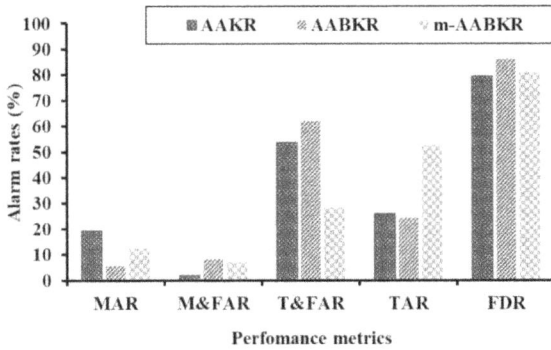

Figure 11.
Alarm rates in a steady-state numerical process.

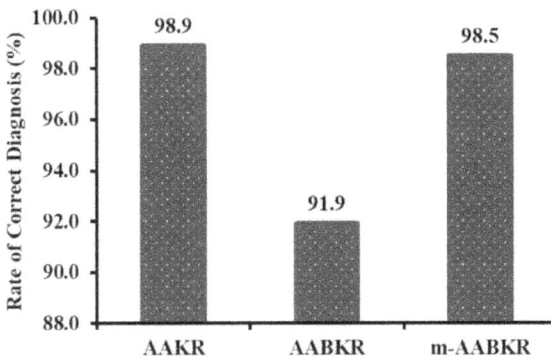

Figure 12.
Rate of correct fault diagnosis in a steady-state numerical process.

observe that the performance of the modified AABKR is comparable to that of AAKR, and the method can, thus, also be used effectively for signal validation during steady-state process operation.

5.2 Start-up transient operations in nuclear power plants

The real-time nuclear simulator data used in [54] is taken here to test the applicability of the proposed methods in transient operations. The data is collected from the simulator without any faults during heating from the cool-down mode (start-up operation). The simulator was designed to reproduce the behavior of a three loop pressurized water reactor (PWR) and to carry out various operational modes, such as start-up, preoperational tests, preheating, hot start-up, cold shut-down, power control, and the operational conditions in steady and accident states. **Figure 13** shows a basic three-loop PWR which is just an illustration of a real process. Six sensors' process signals from the reactor coolant system (RCS) were selected for monitoring during the start-up operation: S1 (cold leg temperature), S2 (core exit temperature), S3 (hot leg temperature), S4 (safety injection flow), S5 (residual heat removal flow), and S6 (sub-cooling margin temperature). The data consist of 1000 observations sequentially collected at constant time intervals of 1 s.

Because these data are fault-free, we first used the entire 1000 observations to train the method and to determine the optimal model parameters using 10-fold cross-validation. Then, we simulated an abnormal condition within these data for use as the testing data set for model evaluation. To effectively examine the fault

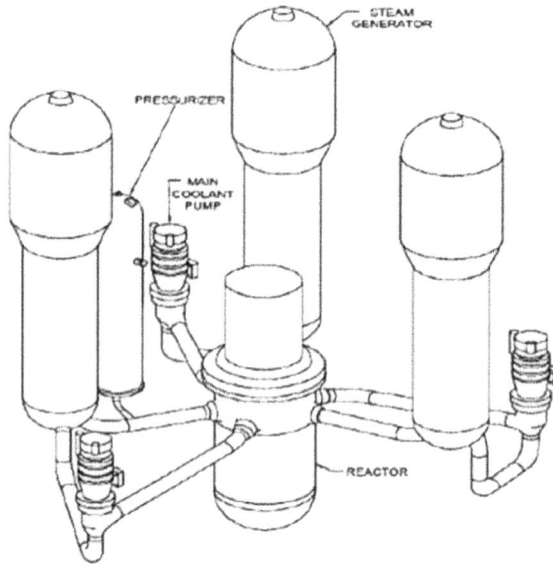

Figure 13.
Schematic diagram of three loops PWR reactor coolant system [60].

detection capability of the proposed method, we conducted a thousand-runs Monte Carlo simulation experiment, in which the fault magnitude was a random number sampled from a bimodal uniform distribution, $U([-10, -2] \cup [2, 10])$, and added to a signal. In this case, the sensor signal in fault at any time step of the Monte Carlo simulation was random uniformly sampled among the six possible sensor signals.

Summary statistics for the alarm rates from AAKR, AABKR, and the modified AABKR from the thousand-runs Monte Carlo simulation experiment are presented in **Table 2**. The mean values of the distributions of the alarm rates are depicted in

Model	Summary statistics	Alarm rates				
		MAR	M&FAR	T&FAR	TAR	FDR
AAKR	Mean	15.43	2.18	21.02	61.37	84.57
	Median	15.41	2.15	21.02	61.35	84.59
	Max	19.54	3.73	25.78	66.13	89.11
	Min	10.89	0.81	16.86	55.82	80.46
AABKR	Mean	0.14	0.01	0.09	99.76	99.86
	Median	0.12	0.00	0.12	99.77	99.88
	Max	0.59	0.23	0.70	100	100
	Min	0.00	0.00	0.00	98.95	99.41
Modified AABKR	Mean	0.01	0.32	4.36	95.30	99.99
	Median	0.00	0.35	4.32	95.33	100
	Max	0.23	0.95	7.50	97.69	100
	Min	0.00	0.00	2.20	91.97	99.76

Table 2.
Alarm rates (%) of a thousand-runs Monte Carlo simulation in start-up data.

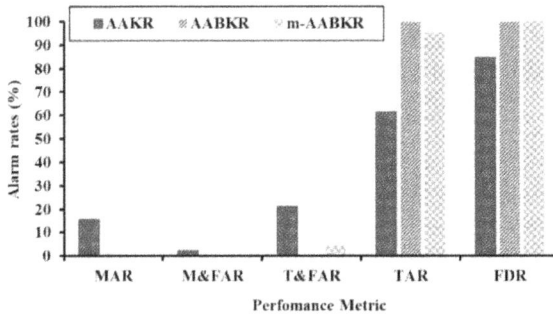

Figure 14.
Means of the alarm rates in start-up process operating condition.

Figure 14. The mean values of the distributions of the TARs are 61.4, 99.8, and 95.3% for AAKR, AABKR, and the modified AABKR, respectively. While the mean values of the distributions of the FDRs are 84.57, 99.86, and 99.99% for AAKR, AABKR, and the modified AABKR, respectively. From the results, the performance of both AABKR and the modified AABKR is better than that of AAKR. On average, the TAR from AABKR is slightly higher than that of the modified AABKR. However, there is no significant difference between the FDR of the AABKR and that of the modified AABKR. Thus, for single-sensor faults, the modified AABKR, on average, has performance similar to that of the AABKR, and can be used to validate the sensors' states during transient operations, with the benefit of eliminating the use of derivatives entirely, thereby extending the applicability to steady-state process operation.

6. Conclusions

In this work, we have considered signal reconstruction models for fault detection in nuclear power plants. In order to improve the performance of the AABKR and extend its applicability to steady-state operating conditions, we have proposed a modification, based on a different procedure for the determination of the time position index, the position of the nearest vector within the memory vector to the query vector observation, which provides the input to the weighted-distance algorithm that captures temporal dependencies in the data. Two different approaches based on DTW for time position index identification, have been developed. The basic idea is that, the use of derivative in AABKR, which becomes constant and nearly zero during steady-state operation when the process change in time is negligible and makes it impossible to identify the time position index correctly, can be completely eliminated while maintaining an acceptable performance in monitoring during both steady-state and transient operations.

The modified AABKR method has been applied, first, to a typical steady-state process and, then, to a case study concerning the monitoring of a reactor coolant system of a PWR NPP during start-up transient operation. We have conducted Monte Carlo simulation experiments to critically examine the fault detection capability of the proposed method and the results have been compared to those of AAKR and AABKR using several performance metrics. The obtained results have shown that the reconstructions provided by the modified AABKR are more robust than those of AAKR and AABKR, in particular, during steady-state operations. The method can, then, be used for signal reconstruction during both steady-state and transient operations, with the benefit of eliminating the use of derivatives entirely while maintaining an acceptable performance. If these approaches are adopted, the cause of abnormalities can be identified, proper maintenance intervention can be planned and earlier mitigation can be allowed to avoid the risk of catastrophic failure.

The future works will focus on (1) the development of an ensemble model in order to benefit from the exploitations of different capabilities of the three methods for signal reconstructions; and (2) the development of a method for on-line updating of the memory data, allowing the model to automatically adapt to the changes in different operating conditions.

A. Appendices

A.1 Auto associative kernel regression

The framework of the AAKR technique comprises three steps, briefly presented below [61, 62].

1) Distance calculation

The distance between the query vector $\mathbf{x}_q$ and each of the memory data vectors is computed. There are many different distance metrics that can be used, but the most commonly used one is the Euclidean distance (L^2-Norm):

$$d_i\left(\mathbf{X}_i, \mathbf{x}_q\right) = \left\|\mathbf{X}_i - \mathbf{x}_q\right\|_2 = \sqrt{\sum_{j=1}^{p}\left(x_{ij} - x_{qj}\right)^2} \tag{A1}$$

For a single query vector, this calculation is repeated for each of M memory vectors, resulting into $\mathbf{d} \in \mathbb{R}^{M \times 1}$.

2) Similarity weight quantification

The distance d_i in vector $\mathbf{d}$ is used to determine the weights for the AAKR, for example, by evaluating the Gaussian kernel:

$$k_i\left(\mathbf{X}_i, \mathbf{x}_q\right) = exp\left(\frac{-d_i^2}{2h^2}\right) \tag{A2}$$

where h is the bandwidth.

3) Output estimation

Finally, the quantified weights (Eq. (A2)) are combined with the memory data vectors to make estimations by using a weighted average:

$$\hat{x}_{qj} = \frac{\sum_{i=1}^{M} k_i x_{ij}}{\sum_{i=1}^{M} k_i} \tag{A3}$$

A.2 Algorithms

A.2.1 Time position index identification—First approach

Algorithm B.1: Time position index identification – First approach

---

Inputs: $\mathbf{X}[M, p]$ and $\mathbf{X}_q^*[r, p]$ /* $\mathbf{X}$ is a matrix of memory time-series data with M observational sequences of p-dimensional signals and $\mathbf{X}_q^*$ is a matrix of query input time-series data of window length, r */
Output: ε /* time position index */

1	**begin**	
2		Generate N matrices from $\mathbf{X}$ with $(r - 1)$ overlapping
3		**for** $k = 1$ to N **do**
4		Get $\mathcal{A}_r^k[r, p]$
5		Compute local cost function matrix, $\mathbf{D}$ using Eq. (32)
6		Compute global cost matrix, $\mathbf{G}$ using Eqs.(27),(28) and (29)
7		Obtain DTW distance using Eq. (31)
8		**end**
9		Determine the time position index, ε using Eq. (33)
10		**return** ε
	end	
11		

A.2.2 Time position index identification—Second approach

Algorithm B.2: Time position index identification – Second approach

---

Inputs: $\mathbf{X}[M, p]$ and $\mathbf{X}_q^*[r, p]$ /* $\mathbf{X}$ is a matrix of memory time-series data with M observational sequences of p-dimensional variables, and $\mathbf{X}_q^*$ is a matrix of query input time-series data of window length, r */
Output: ε /* time position index */

1	**begin**	
2		Compute local cost function matrix, $\mathbf{D}$ using Eq. (32)
3		Compute global cost matrix $\mathbf{G}$ using Eqs.(34),(35) and (36)
4		Get the last row of $\mathbf{G}$ as $\mathbf{G}[r,]$
5		Determine the time position index, ε using Eq. (37)
6		**return** ε
	end	
7		

A.3 Performance metrics

We evaluated and compared the proposed methods using a set of performance metrics proposed in [54]: missed alarm rate (MAR), missed and false alarms rate (M&FAR), true and false alarms rate (T&FAR), true alarm rate (TAR), and fault detection rate (FDR). These metrics are briefly summarized as follows:

A.3.1 Missed alarm rate

A missed alarm occurs when at least one process variable is erroneously not detected as faulty, $\left(|e_j| \leq T_j^D \right)$, when a fault is actually present. In this case, at least one missed alarm occurs, and no false alarm occurs in any of the other variables. The MAR is calculated as:

$$\text{MAR} = \frac{\sum missed\ alarms}{total\ number\ of\ samples\ in\ fault\ condition} * 100\% \qquad (A4)$$

A.3.2 Missed and false alarms rate

It is possible that a fault will be missed in a faulty signal but detected in at least one fault-free signal. This gives both missed and false alarms: a fault is detected $\left(|e_j| > T_j^D \right)$ in at least one process signal, when no fault is actually present (false alarm), and at least one process signal has a fault that is not detected $\left(|e_j| \leq T_j^D \right)$ (missed alarm). The M&FAR is calculated as:

$$\text{M\&FAR} = \frac{\sum simultaneous\ missed\&false\ alarms}{total\ number\ of\ samples\ in\ fault\ condition} * 100\% \qquad (A5)$$

A.3.3 True and false alarms rate

It is possible that a fault will be detected in a faulty signal and also detected in at least one fault-free signal. This gives both true and false alarms: a fault is detected in at least one process signal, $\left(|e_j| > T_j^D \right)$, when no fault is actually present (false alarm), and a fault is correctly detected in one process signal, $\left(|e_j| > T_j^D \right)$ (true alarm). The T&FAR is calculated as:

$$\text{T\&FAR} = \frac{\sum simultaneous\ true\&false\ alarms}{total\ number\ of\ samples\ in\ fault\ condition} * 100\% \qquad (A6)$$

A.3.4 True alarm rate

This represents the presence of only true alarms. A fault is detected in at least one process signal, $\left(|e_j| > T_j^D \right)$, when a fault is actually present, and no false alarm exists in other fault-free signals (true alarm). The TAR is calculated as:

$$\text{TAR} = \frac{\sum true\ alarms|no\ false\ alarms}{total\ number\ of\ samples\ in\ fault\ condition} * 100\% \qquad (A7)$$

A.3.5 Fault detection rate

The FDR is expressed as the ratio of the number of faulty data points detected as faulty to the total number of samples specific to a fault. In this case, a fault is detected in at least one process signal, $\left(|e_j| > T_j^D \right)$, when a fault is actually present regardless of false alarms in other signals. The FDR is calculated as:

$$\text{FDR} = \frac{\sum correctly\ decteted\ faults\ in\ the\ system}{total\ number\ of\ samples\ in\ fault\ condition} * 100\% \qquad (A8)$$

FDR measures the ability of a model to detect the presence of the fault in a system when a fault is actually present. Thus, FDR is a summation of the M&FAR, T&FAR, and TAR, which implies that:

$$FDR = M\&FAR + T\&FAR + TAR \qquad (A9)$$

Nomenclature

Abbreviations

AABKR	auto-associative bilateral KR
AAKR	auto-associative KR
AANN	auto-associative artificial neural network
DTW	dynamic time warping
ECM	evolving clustering method
FDD	fault detection and diagnosis
FDR	fault detection rate
GPR	Gaussian process regression
KR	kernel regression
M&FAR	missed and false alarms rate
MAR	missed alarm rate
MSET	multivariate state estimation technique
NPPs	nuclear power plants
PCA	principal component analysis
PLS	partial least squares
PWR	pressurized water reactor
RCS	reactor coolant system
RMSE	root mean square error
SSCs	systems, structures, and components
SVM	support vector machine
T&FAR	true and false alarms rate
TAR	true alarm rate

Symbols and notations

$\mathbf{X}$	matrix of training/memory data set
$\mathbf{X}_q^*$	query test pattern $r \times p$ matrix
$\mathbf{x}_{q_r}^*$	the last rth vector in $\mathbf{X}_q^*$
$\mathcal{A}$	array containing N matrices generated from $\mathbf{X}$
A_r^k	kth $r \times p$ matrix in the array $\mathcal{A}$
M	number of observations in the memory data
N	number of matrices in the array $\mathcal{A}$ generated from $\mathbf{X}$
p	number of process variables
r	sliding window length
i	observation time index of memory data
q	a subscript, indicating query input
j	process variable index
x_{ij}	ith observation of the jth variable
t	time, independent variable
$\hat{x}_{q_r}$	estimated value of $\mathbf{x}_{q_r}^*$
d	distance metric
k	kernel weight of AAKR
h	kernel bandwidth of AAKR
t_q	query time input
f	feature component of AABKR

$\mathbf{k}^f$	feature kernel vector
$\mathbf{k}^t$	temporal kernel vector
h_f	feature kernel bandwidth
h_t	temporal kernel bandwidth
η	constant time interval
ε	time position index
$\partial \mathcal{A}_r^k / \partial t$	derivative of the last row in $\mathcal{A}_r^k$ with respect to time, t
$\partial \mathbf{x}_{q_r}^* / \partial t$	derivative of current measurement vector in $\mathbf{X}_q^*$ with respect to t
$\boldsymbol{\Delta}$	derivative distance vector
$\boldsymbol{\delta}$	temporal weighted distance vector
$\mathbf{k}^{ab}$	adaptive bilateral kernel vector
$\mathbf{e}_{q_r}$	residual vector
T_j^D	threshold for fault detection in the jth variable
$\mathbf{D}$	local cost (distance) matrix of the DTW
$\mathbf{G}$	global cost (accumulated) matrix of the DTW

Author details

Ibrahim Ahmed[1], Enrico Zio[1,2*] and Gyunyoung Heo[3]

1 Department of Energy, Politecnico di Milano, Milan, Italy

2 MINES ParisTech, PSL Research University, CRC, Sophia Antipolis, France

3 Department of Nuclear Engineering, Kyung Hee University, Yongin-si, Republic of Korea

*Address all correspondence to: enrico.zio@polimi.it

IntechOpen

References

[1] Rouhiainen V. Safety and Reliability: Technology Theme—Final Report. Espoo: VTT Publications 592; 2006

[2] Coble J, Ramuhalli P, Bond L, Hines JW, Upadhyaya B. A review of prognostics and health management applications in nuclear power plants. International Journal of Prognostics and Health Management. 2015;**6**:1-22

[3] IAEA. On-Line Monitoring for Improving Performance of Nuclear Power Plants Part 2: Process and Component Condition Monitoring and Diagnostics. IAEA Nuclear Energy Series: No. NP-T-1.2. Vienna, Austria: IAEA; 2008

[4] EPRI. On-Line Monitoring of Instrument Channel Performance, Volume 1: Guidelines for Model Development and Implementation. Vol. Vol. 1. Palo Alto, CA: EPRI; 2004. p. 1003361

[5] IAEA. On-Line Monitoring for Improving Performance of Nuclear Power Plants Part 1: Instrument Channel Monitoring. IAEA Nuclear Energy Series: No. NP-T-1.1. Vienna, Austria: IAEA; 2008

[6] EPRI. Equipment Condition Assessment Volume 1: Application of On-Line Monitoring Technology. Vol. Vol. 1. Palo Alto, CA: EPRI; 2004. p. 1003695

[7] Heo G, Lee SK. Internal leakage detection for feedwater heaters in power plants using neural networks. Expert Systems with Applications. 2012; **39**:5078-5086

[8] An SH, Heo G, Chang SH. Detection of process anomalies using an improved statistical learning framework. Expert Systems with Applications. 2011;**38**: 1356-1363

[9] Li F, Upadhyaya BR, Coffey LA. Model-based monitoring and fault diagnosis of fossil power plant process units using group method of data handling. ISA Transactions. 2009;**48**: 213-219

[10] Isermann R. Fault-Diagnosis Applications: Model-Based Condition Monitoring—Actuators, Drives, Machinery, Plants, Sensors, and Fault-Tolerant Systems. Berlin Heidelberg, Springer; 2011

[11] Nabeshima K, Suzudo T, Suzuki K. Real-time nuclear power plant monitoring with neural network. Journal of Nuclear Science and Technology. 1998;**35**:93-100

[12] Baraldi P, Gola G, Zio E, Roverso D, Hoffmann M. A randomized model ensemble approach for reconstructing signals from faulty sensors. Expert Systems with Applications. 2011;**38**: 9211-9224

[13] Jiang Q, Yan X. Plant-wide process monitoring based on mutual information-multiblock principal component analysis. ISA Transactions. 2014;**53**:1516-1527

[14] Ding SX. Model-Based Fault Diagnosis Techniques: Design Schemes, Algorithms, and Tools. Berlin Heidelberg, Springer; 2008

[15] Chen J. Robust residual generation for model-based fault diagnosis of dynamic systems [Ph.D. thesis]. University of York; 1995

[16] Patton RJ, Chen J. Observer-based fault detection and isolation: Robustness and applications. Control Engineering Practice. 1997;**5**:671-682

[17] Simani S, Fantuzzi C, Patton RJ. Model-Based Fault Diagnosis in

Dynamic Systems Using Identification Techniques. London, Springer; 2002

[18] Chiang LH, Russell EL, Braatz RD. Fault Detection and Diagnosis in Industrial Systems. London, Springer; 2000

[19] Ma J, Jiang J. Applications of fault detection and diagnosis methods in nuclear power plants: A review. Progress in Nuclear Energy. 2011;53: 255-266

[20] Garvey J, Garvey D, Seibert R, Hines JW. Validation of on-line monitoring techniques to nuclear plant data. Nuclear Engineering and Technology. 2007;39:149-158

[21] Zio E. Prognostics and health management of industrial equipment. In: Kadry S, editor. Diagnostics and Prognostics of Engineering Systems: Methods and Techniques. USA: IGI Global; 2012. pp. 333-356

[22] Zio E, Di Maio F, Stasi M. A data-driven approach for predicting failure scenarios in nuclear systems. Annals of Nuclear Energy. 2010;37:482-491. DOI: 10.1016/j.anucene.2010.01.017

[23] EPRI. On-Line Monitoring for Equipment Condition Assessment: Application at Progress Energy. Palo Alto, CA: EPRI; 2004. p. 1008416

[24] Ahmad R, Kamaruddin S. An overview of time-based and condition-based maintenance in industrial application. Computers and Industrial Engineering. 2012;63:135-149

[25] Zio E, Broggi M, Pedroni N. Nuclear reactor dynamics on-line estimation by locally recurrent neural networks. Progress in Nuclear Energy. 2009;51: 573-581. DOI: 10.1016/j. pnucene.2008.11.006

[26] Simani S, Marangon F, Fantuzzi C. Fault diagnosis in a power plant using

artificial neural networks: Analysis and comparison. In: European Control Conference. Karlsruhe, Germany: IEEE; 1999. pp. 2270-2275

[27] Na MG, Shin SH, Jung DW, Kim SP, Jeong JH, Lee BC. Estimation of break location and size for loss of coolant accidents using neural networks. Nuclear Engineering and Design. 2004; 232:289-300

[28] Yao M, Wang H. On-line monitoring of batch processes using generalized additive kernel principal component analysis. Journal of Process Control. 2015;28:56-72

[29] Dunia R, Qin SJ, Edgar TF, McAvoy TJ. Identification of faulty sensors using principal component analysis. AICHE Journal. 1996;42: 2797-2812

[30] Harkat M-F, Djelel S, Doghmane N, Benouaret M. Sensor fault detection, isolation and reconstruction using nonlinear principal component analysis. International Journal of Automation and Computing. 2007;4:149-155

[31] Heo G, Choi SS, Chang SH. Thermal power estimation by fouling phenomena compensation using wavelet and principal component analysis. Nuclear Engineering and Design. 2000;199: 31-40

[32] Cui P, Li J, Wang G. Improved kernel principal component analysis for fault detection. Expert Systems with Applications. 2008;34:1210-1219

[33] Zhao C, Gao F. Fault-relevant principal component analysis (FPCA) method for multivariate statistical modeling and process monitoring. Chemometrics and Intelligent Laboratory Systems. 2014;133:1-16

[34] Gross KC, Singer RM, Wegerich SW, Herzog JP, VanAlstine R, Bockhorst F. Application of a model-

based fault detection system to nuclear plant signals. In: International Conference on Intelligent Systems Applications to Power Systems (ISAP'97). Seoul, Korea, Argonne National Laboratory; 1997

[35] Singer RM, Gross KC, Herzog JP, King RW, Wegerich S. Model-based nuclear power plant monitoring and fault detection: Theoretical foundations. In: International Conference on Intelligent Systems Applications to Power Systems (ISAP'97). Seoul, Korea, Argonne National Laboratory; 1997

[36] Chen X, Xu G. A self-adaptive alarm method for tool condition monitoring based on Parzen window estimation. Journal of Vibroengineering. 2013;15: 1537-1545

[37] Mandal SK, Chan FTS, Tiwari MK. Leak detection of pipeline: An integrated approach of rough set theory and artificial bee colony trained SVM. Expert Systems with Applications. 2012; 39:3071-3080

[38] Rocco CM, Zio E. A support vector machine integrated system for the classification of operation anomalies in nuclear components and systems. Reliability Engineering and System Safety. 2007;92:593-600

[39] Zio E, Baraldi P, Zhao W. Confidence in signal reconstruction by the evolving clustering method. In: Proceedings of the Annual Conference of the Prognostics and System Health Management (PHM 2011). Shenzhen: IEEE; 2011

[40] Muradore R, Fiorini P. A PLS-based statistical approach for fault detection and isolation of robotic manipulators. IEEE Transactions on Industrial Electronics. 2012;59:3167-3175

[41] Zio E, Di Maio F. A fuzzy similarity-based method for failure detection and recovery time estimation. International

Journal of Performability Engineering. 2010;6:407-424

[42] Zio E, Di Maio F. A data-driven fuzzy approach for predicting the remaining useful life in dynamic failure scenarios of a nuclear system. Reliability Engineering and System Safety. 2010;95:49-57. DOI: 10.1016/j.ress.2009.08.001

[43] Zio E, Gola G. A neuro-fuzzy technique for fault diagnosis and its application to rotating machinery. Reliability Engineering and System Safety. 2009;94:78-88. DOI: 10.1016/j. ress.2007.03.040

[44] Zio E, Baraldi P, Popescu IC. A fuzzy decision tree method for fault classification in the steam generator of a pressurized water reactor. Annals of Nuclear Energy. 2009;36:1159-1169. DOI: 10.1016/j.anucene.2009.04.011

[45] Marseguerra M, Zio E, Baraldi P, Oldrini A. Fuzzy logic for signal prediction in nuclear systems. Progress in Nuclear Energy. 2003;43:373-380

[46] Ding SX. Data-driven design of monitoring and diagnosis systems for dynamic processes: A review of subspace technique based schemes and some recent results. Journal of Process Control. 2014;24:431-449

[47] Hu Y, Palmé T, Fink O. Fault detection based on signal reconstruction with auto-associative extreme learning machines. Engineering Applications of Artificial Intelligence. 2017;57:105-117

[48] Hines JW, Garvey J, Garvey DR, Seibert R. Technical Review of On-Line Monitoring Techniques for Performance Assessment Volume 3: Limiting Case Studies. U.S. Nuclear Regulatory Commission, NUREG/CR-6895. Vol. 32008. Washington DC, USA, US Nuclear Regulatory Commission

[49] Baraldi P, Di Maio F, Genini D, Zio E. Comparison of data-driven

reconstruction methods for fault detection. IEEE Transactions on Reliability. 2015;**64**:852-860

[50] Hines JW. Robust distance measures for on-line monitoring. US Patent. Patent No.: US008311774 B2; 2012

[51] Baraldi P, Di Maio F, Turati P, Zio E. Robust signal reconstruction for condition monitoring of industrial components via a modified auto associative kernel regression method. Mechanical Systems and Signal Processing. 2015;**60**:29-44

[52] Ahmed I, Heo G. Development of a modified kernel regression model for a robust signal reconstruction. In: Transactions of the Korean Nuclear Society Virtual Autumn Meeting. Gyeongju, Korea, Korean Nuclear Society; 2016

[53] Ahmed I, Heo G. Development of a transient signal validation technique via a modified kernel regression model. In: 10th International Embedded Topical Meeting on Nuclear Plant Instrumentation, Control and Human-Machine Interface Technologies (NPIC&HMIT 2017). San Francisco, CA, USA, American Nuclear Society; 2017. pp. 1943-1951

[54] Ahmed I, Heo G, Zio E. On-line process monitoring during transient operations using weighted distance auto associative bilateral kernel regression. ISA Transactions. 2019;**92**:191-212. DOI: 10.1016/j.isatra.2019.02.010

[55] Ahmed I. Bilateral kernel methods for time-series states validation in process systems [Ph.D. thesis]. Republic of Korea: Department of Nuclear Engineering, Kyung Hee University; 2020

[56] Sakoe H, Chiba S. Dynamic programming algorithm optimization for spoken word recognition. IEEE Transactions on Acoustics, Speech, and Signal Processing. 1978;**26**:43-49

[57] Meinard M. Dynamic Time Warping. Information Retrieval for Music and Motion. Berlin, Heidelberg: Springer; 2007. DOI: 10.1007/978-3-540-74048-3_4

[58] Berndt DJ, Clifford J. Using dynamic time warping to find patterns in time series. In: Work Knowledge Discovery in Databases. Palo Alto, California, USA, AAAI; 1994. pp. 359-370

[59] Alcala CF, Qin SJ. Reconstruction-based contribution for process monitoring. Automatica. 2009;**45**: 1593-1600

[60] Xiao B-B, Wang T-C. An investigation of FLEX implementation in Maanshan NPP by using MAAP 5. In: Jiang H, editor. Proceedings of the 20th Pacific Basin Nuclear Conference (PBNC 2016). Singapore: Springer; 2017. pp. 81-96. DOI: 10.1007/978-981-10-2311-8_8

[61] Heo G. Condition monitoring using empirical models: Technical review and prospects for nuclear applications. Nuclear Engineering and Technology. 2008;**40**:49-68

[62] Hines JW, Garvey D, Seibert R, Usynin A. Technical Review of On-Line Monitoring Techniques for Performance Assessment, Volume 2: Theoretical Issues. U.S. Nuclear Regulatory Commission, NUREG/CR-6895. Vol. Vol. 22008. Washington DC, USA, US Nuclear Regulatory Commission

www.ingramcontent.com/pod-product-compliance
Lightning Source LLC
Chambersburg PA
CBHW081544190326
41458CB00015B/5633